CPSS Power Electronics Series

This series comprises advanced textbooks, research monographs, professional books, and reference works covering different aspects of power electronics, such as Variable Frequency Power Supply, DC Power Supply, Magnetic Technology, New Energy Power Conversion, Electromagnetic Compatibility as well as Wireless Power Transfer Technology and Equipment. The series features leading Chinese scholars and researchers and publishes authored books as well as edited compilations. It aims to provide critical reviews of important subjects in the field, publish new discoveries and significant progress that has been made in development of applications and the advancement of principles, theories and designs, and report cutting-edge research and relevant technologies. The CPSS Power Electronics series has an editorial board with members from the China Power Supply Society and a consulting editor from Springer.

Readership: Research scientists in universities, research institutions and the industry, graduate students, and senior undergraduates.

Indexed by EI Compendex.

Yijie Wang · Jianwei Mai · Peng Gu · Dianguo Xu

Long Range Inductive Power Transfer

Springer

Yijie Wang
Harbin Institute of Technology
Harbin, Heilongjiang, China

Jianwei Mai
Harbin Institute of Technology
Harbin, Heilongjiang, China

Peng Gu
Northeastern University
Shenyang, Liaoning, China

Dianguo Xu
Harbin Institute of Technology
Harbin, Heilongjiang, China

ISSN 2520-8853 ISSN 2520-8861 (electronic)
CPSS Power Electronics Series
ISBN 978-981-96-7540-1 ISBN 978-981-96-7541-8 (eBook)
https://doi.org/10.1007/978-981-96-7541-8

This Springer imprint is published by the registered company Springer Nature Singapore Pte Ltd.
The registered company address is: 152 Beach Road, #21-01/04 Gateway East, Singapore 189721, Singapore

Preface

The rapid advancement of wireless power transfer (WPT) technologies has ushered in transformative possibilities across industries, from smart grids and industrial automation to electric vehicle charging and biomedical devices. Among these innovations, inductive power transfer (IPT) stands out for its ability to deliver energy across air gaps without physical contact. However, extending the operational range of IPT systems while maintaining efficiency, reliability, and practicality remains a formidable challenge. This book, Long-Range Inductive Power Transfer, addresses these challenges head-on, offering a comprehensive exploration of theoretical breakthroughs, design methodologies, and practical implementations that push the boundaries of wireless energy delivery.

Traditional IPT systems excel in short-range applications but face significant limitations as transmission distances increase. Efficiency degradation, magnetic coupling instability, parasitic effects, and load variability emerge as critical barriers. These challenges are particularly acute in emerging applications such as high-voltage infrastructure monitoring, automated industrial robotics, and dynamic charging for electric vehicles, where extended operational ranges and misalignment tolerance are paramount. This book responds to these demands by dissecting the core issues and presenting innovative solutions that harmonize theoretical rigor with engineering practicality.

This book includes nine chapters.

Chapter 1 introduces the principles of long-range IPT, emphasizing its applications and challenges. It critically evaluates planar and three-dimensional magnetic coupling structures, highlighting their trade-offs in efficiency and misalignment tolerance. The impact of coil parasitic parameters and rectifier impedance characteristics is rigorously analyzed, setting the stage for subsequent solutions.

Chapter 2 delves into compensation topologies, the backbone of resonant IPT systems. By comparing S/S, LCC/S, LCC/LCC, and multi-stage configurations, this chapter provides a roadmap for optimizing impedance matching and minimizing reactive power losses in loosely coupled transformers.

Chapter 3 tackles the often-overlooked issue of coil quality factor degradation due to parasitic capacitance. Through distributed parameter modeling and novel segmented compensation strategies, this section reveals how to enhance coil efficiency without compromising compactness.

Chapter 4 pioneers the design of three-dimensional magnetic couplers, introducing the distance-to-diameter ratio (DDR) as a universal metric for evaluating long-range performance. The cubic magnetic coupler (CMC) emerges as a breakthrough, balancing high coupling coefficients with exceptional misalignment tolerance.

Chapter 5 extends these principles to multi-stage systems, proposing a double-layer compensation architecture that achieves unprecedented transmission distances (2.5 m) while maintaining efficiency. Experimental validation underscores its superiority over conventional parallel and series topologies.

Chapter 6 confronts the reactive power challenges inherent in rectifier circuits. By reimagining C-filter and LC-filter rectifiers, this chapter demonstrates how adaptive compensation and topology modifications can expand operational ranges and stabilize system dynamics.

Target Audience

This book is designed for researchers, engineers, and graduate students in power electronics, electrical engineering, and applied physics. It bridges the gap between theoretical models and real-world implementations, offering actionable insights for designing next-generation WPT systems. Practitioners will find value in the experimental validations and design guidelines, while academics will appreciate the rigorous analytical frameworks and innovative metrics like distance-to-diameter ratio.

Harbin, China

Yijie Wang
Jianwei Mai
Peng Gu
Dianguo Xu

Acknowledgments This research monograph systematically summarizes the research work on the control techniques for long-range inductive power transfer since 2014.

The work in this book was partially supported by the National Natural Science Foundation of China (Grants 51922033 and 52407198), as well as the Delta Power Electronics Science and Education Development Foundation under Award DREK2020003. We sincerely appreciate their generous support and contributions to this research.

It has been an immense pleasure collaborating with the colleagues from Springer, Science Press, and the China Power Supply Society (CPSS).

Competing Interests The authors have no competing interests to declare that are relevant to the content of this manuscript.

Contents

About the Authors

Yijie Wang was born in Heilongjiang Province, China, in 1982. He received the B.S., M.S., and Ph.D. degrees in electrical engineering from Harbin Institute of Technology, China, in 2005, 2007, and 2012, respectively.

From 2012 to 2014, he was a lecturer with the Department of Electrical and Electronics Engineering, Harbin Institute of Technology. From 2014 to 2017, he was an associate professor with the Department of Electrical and Electronics Engineering, Harbin Institute of Technology. Since 2017, he has been a professor with the Department of Electrical and Electronics Engineering, Harbin Institute of Technology. His interests include DC-DC converters, soft-switching power converters, power factor correction circuits, digital control electronic ballasts, and LED lighting systems.

Dr. Wang is an Associate Editor of *IEEE Transactions on Industrial Electronics*, *IEEE Journal of Emerging and Selected Topics in Power Electronics*, *IEEE Access*, *IET Power Electronics*, and *Journal of Power Electronics*.

Jianwei Mai was born in Henan, China, in 1994. He received the B.S. and Ph.D. degrees in electrical engineering from Harbin Institute of Technology, China, in 2017 and 2022, respectively.

From 2023 to 2024, he was an assistant researcher at the Department of Electrical and Electronics Engineering, Harbin Institute of Technology. From 2024 to 2025, he was an assistant professor; since 2025, he has been an associate researcher. His research interests include inductive power transfer, magnetic coupling structure design and simultaneous wireless power, and data transfer.

Dr. Mai is an Associate Editor of *Wireless Power Transfer* and a Guest Associate Editor of *IEEE Journal of Emerging and Selected Topics in Power Electronics.*

Peng Gu was born in Harbin, China. He received the B.S. degree from Huazhong University of Science and Technology, Wuhan, China, in 2015, the M.S. degree from China Electric Power Research Institute, Beijing, China, in 2018, and the Ph.D. degree from Harbin Institute of Technology, Harbin, China, in 2023, all in electrical engineering.

Dr. Gu is currently working as a faculty member in Northeastern University, Shenyang, China, in electrical engineering, since 2023. He is also working as a postdoctoral fellow in Northeastern University in control science and engineering, since 2023. His current research interests include wireless power transfer, magnetic coupling structure design, power system analysis, electromagnetic transient analysis, and DC-DC converter.

Dianguo Xu (Fellow, IEEE) received the B.S. degree in control engineering from Harbin Engineering University, Harbin, China, in 1982, and the M.S. and Ph.D. degrees in electrical engineering from the Harbin Institute of Technology (HIT), Harbin, China, in 1984 and 1989, respectively.

In 1984, he joined the Department of Electrical Engineering, HIT, as an Assistant Professor. Since 1994, he has been a Professor with the Department of Electrical Engineering, HIT. From 2000 to 2010, he was the Dean of the School of Electrical Engineering and Automation, HIT. He was the Vice President of HIT from 2014 to 2020. He authored/coauthored more than 600 technical papers. His research interests include renewable energy generation technology, power quality mitigation, sensorless vector controlled motor drives, and high-performance servo system.

Dr. Xu is a Chairperson of IEEE Harbin Section, Co-EIC of *IEEE Transactions on Power Electronics*, and an Associate Editor for the *IEEE Transactions on Industrial Electronics* and *IEEE Journal of Emerging and Selected Topics in Power Electronics*. He was the recipient of the 2018 IEEE IAS Outstanding Achievement Award.

Chapter 1
Introduction

Abstract This chapter provides a comprehensive overview of inductive power transfer (IPT) technology, focusing on its application in long-range wireless power transmission (WPT). It begins by exploring the growing demand for WPT in high-voltage monitoring, smart grids, and industrial automation, emphasizing challenges such as efficiency degradation, magnetic coupling limitations, and parasitic effects at extended distances. The analysis systematically reviews magnetic coupling structures, including planar and three-dimensional designs, evaluating their trade-offs between misalignment tolerance, coupling efficiency, and air-gap extension. Critical discussions address the impact of coil parasitic parameters and rectifier impedance characteristics on system performance, alongside strategies like distributed compensation and harmonic mitigation to enhance efficiency. Emerging solutions, such as optimized core geometries and adaptive impedance matching, are highlighted as pathways to overcome existing barriers. The chapter concludes with future directions for advancing WPT systems, aiming to bridge theoretical insights with practical scalability for next-generation applications.

Keywords Inductive power transfer (IPT) · Magnetic coupling structures · Long-range transmission · Parasitic parameters · Rectifier impedance

1.1 Application of Inductive Power Transfer with Long Range

Wireless Power Transfer (WPT) technology has the advantages of strong environmental adaptability, high degree of freedom in energy transmission, good safety, and electrical isolation [1, 2]. It is widely used in electric vehicles and drones, rail transportation, underwater equipment, consumer electronics, biomedicine and other fields have broad application prospects. WPT technologies mainly include electromagnetic induction, electric field coupling, electromagnetic radiation (microwave, laser), and ultrasonic methods [3]. Among them, the research level of electromagnetic induction wireless power transfer (IPT) technology based on near-field coupling is the most

Y. Wang et al., *Long Range Inductive Power Transfer*, CPSS Power Electronics Series,
https://doi.org/10.1007/978-981-96-7541-8_1

mature. Compared with the other three wireless energy transmission technologies, the power and efficiency level of its energy transmission can reach a relatively high level and has achieved specific applications in related industrial fields.

Nikola Tesla, a pioneer in the field of electrical engineering, was the main pioneer of wireless energy transmission technology. At the end of the nineteenth century, Tesla proposed a great idea of achieving global wireless energy transmission using the earth itself through theoretical analysis and modeling [4]. With the continuous development of power electronics technology in the twentieth century, WPT has gradually become possible with the help of power converter models. Since the 1990s, the scientific research team of the University of Auckland in New Zealand has conducted long-term and extensive research on near-field inductively coupled power transfer technology, has made important progress in the fields of magnetic coupling structures, circuit and system topology, and control solutions, and promotes the commercial application of wireless charging technology for electric vehicles [5, 6]. In 2007, a scientific research team from the Massachusetts Institute of Technology published an article in Science magazine [7], using a 60 cm diameter power transmission coil, in the 9.9 MHz electromagnetic coupling resonance system, the radio energy transmission with a distance of 2 m, and an efficiency of about 40% have been realized, which indicates that the development of the radio energy transmission technology has entered a new era.

With the development of WPT technology and the continuous expansion of its application scenarios, people's demand for further extending the WPT distance is increasing [8]. However, in existing research, in order to achieve meter-level WPT, the magnetic coupling structures designed by researchers often have problems such as being too large and too low in efficiency [9, 10], which results in the limitation of the relative distance of WPT although the absolute distance is extended. Some scholars have proposed using a magnetic coupling structure composed of domino planar coils to achieve long air gap WPT [11, 12]. This system usually requires more than ten coils to achieve meter-level WPT distance. For longer wireless distance, it severely limits the efficiency and reliability of the system. Some scholars have achieved high-frequency, long-distance WPT by increasing the frequency to the MHz or GHz level [13, 14]. However, due to the limitations of ultra-high frequency converter technology, switching elements of WPT systems often produce large voltage stress. In addition, the power transmission level of high-frequency WPT technology is low, and its application scenarios are also limited.

In order to promote the intelligentization and digitalization process of the power system and sense the working status of high-voltage electrical equipment in real time, more and more monitoring and transmission systems are used to promote the intelligentization and digitalization process of the power system and sense the working status of high-voltage electrical equipment in real time. More and more monitoring and sensing devices, such as sensors, cameras, and data concentrators, are being applied to the high-voltage side of the power system to sense the operation of high-voltage equipment in real time [15]. However, due to the high voltage difference between the primary side of the power system and the ground, it is impossible to directly provide power to these devices through wire transmission. In order to solve the above problems, some studies have proposed the use of current transformers, photovoltaic panels, lithium battery energy storage, and other solutions to power secondary equipment [16]. However, due to the instability of line power flow, uncertainty of weather, limitations of lithium battery capacity, and other factors, the above scheme has certain limitations. In response to the above problems, WPT schemes with advantages such as safety, stability, and controllability have been successively proposed [17]. For example, integrating a wireless power transmission system with high-voltage insulators for sensor power supply, as shown in Fig. 1.1.

Long air-gap Wireless Power Transfer (WPT) technology has demonstrated substantial potential across multidisciplinary applications. In aerospace engineering, this technology facilitates extended mission durations for unmanned aerial vehicles (UAVs) and satellites through remote energy replenishment systems, effectively reducing operational downtime associated with conventional refueling cycles. Within industrial automation frameworks, long air-gap WPT enables persistent non-contact power delivery to automated guided vehicles (AGVs) and mobile robotic systems, thereby optimizing production line flexibility while eliminating maintenance interruptions caused by physical connectors. The residential automation sector benefits from room-spanning wireless energy distribution architectures that support smart home ecosystems, including intelligent appliances and IoT-enabled lighting networks. Marine engineering applications particularly highlight its capability to overcome depth constraints through subsea WPT implementations, providing reliable energy transmission to benthic monitoring equipment without cable dependency. Recent advancements in magnetically coupled resonance mechanisms and

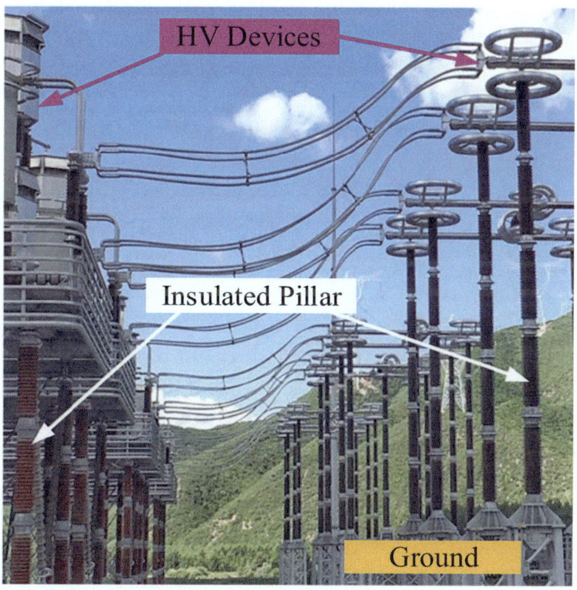

Fig. 1.1 Application scenarios of post insulators in substations

adaptive impedance matching techniques have further enhanced transmission efficiency across extended air gaps, suggesting promising avenues for implementing this technology in smart grid infrastructures and distributed sensor networks. These developments underscore the transformative potential of long air-gap WPT in advancing next-generation wireless energy systems.

1.2 Magnetic Coupling Structures

Compared with the conventional DC-DC isolated resonant converter, the main difference of the inductively coupled power transfer system is that the primary and secondary coupling mechanisms of the transformer part are independent of each other, and there is a certain air gap, which constitutes the core of energy transmission in WPT system. The coupler is called loosely coupled transformer. For different working conditions and application scenarios, the magnetic circuit requirements are different. Therefore, the analysis and design of the magnetic coupling structure of loose coupling transformers is the main direction of WPT technology in the academic research field.

1.2.1 Planar Shape Magnetic Coupling Structure

IPT usually uses a planar magnetic coupling structure to transmit energy, especially in the field of wireless charging of electric vehicles where this technology is most widely used and has the greatest commercial potential.

Among them, the planar circular magnetic coupling structure shown in Fig. 1.2a and the planar square magnetic coupling structure shown in Fig. 1.2b are the most widely used types of magnetic coupling structures. The three-dimensional finite element analysis model was adopted by Budhia et al. [5] to optimize the design of the planar circular magnetic coupling structure, which effectively reduced the weight of the magnetic structure. The mutual inductance of planar circular magnetic coupling structures was analyzed in detail by analytical modeling by Acero et al. [18]. The parameters of the planar circular magnetic structure were optimized and designed based on intelligent algorithms through the multi-objective hybrid particle swarm optimization algorithm and the multi-objective real particle swarm optimization algorithm by Yilmaz et al. [19], which promoted the intelligent design of magnetic coupling structure parameters. Similar to the planar circular magnetic coupling structure, some scholars have proposed a planar square magnetic coupling structure and achieved high-efficiency WPT at the kilowatt level [20]. Some scholars have optimized the design of a WPT system based on a planar square magnetic coupling structure through an asymmetric design method and obtained an improved solution for wireless charging scenarios of electric vehicles [21]. In addition, based on the geometric characteristics and magnetic field distribution of planar circular and planar square magnetic coupling structures, the two most widely used magnetic coupling structures for WPT were analytically modeled and comparatively analyzed, and corresponding design strategies were given for different applicable situations [22, 23]. The geometric shapes of the above-mentioned planar circular and planar square magnetic coupling structures are simple and clear and are easy for industrial production. However, it is difficult to maintain its coupling characteristics when the position of the single-side coil is shifted, and it does not have obvious misalignment tolerance capability.

Facing the field of wireless charging of electric vehicles, the static WPT system has a certain degree of randomness in the spatial relative positions of its primary and secondary side magnetic coupling structures depending on the car's parking position, resulting in the condition of misalignment (offset) between coils. Therefore, it is necessary to conduct research on the anti-offset performance of the magnetic coupling structure of WPT technology. The planar solenoid magnetic coupling structure shown in Fig. 1.2c was proposed by scholars at Saitama University in Japan [24]. Due to the divergence and constraint effect of its magnetic field, it has strong misalignment tolerance capability in a single direction. In the subsequent development process, the planar solenoid magnetic coupling structure was optimized and improved by scholars, which greatly reduced its weight and cost while maintaining its coupling coefficient and excellent anti-offset capability [25]. The DD-shaped magnetic coupling structure shown in Fig. 1.2d was proposed by a research team from the University of Auckland in New Zealand [26], which not only restrains the leakage magnetic field of the magnetic coupling structure, but also improves

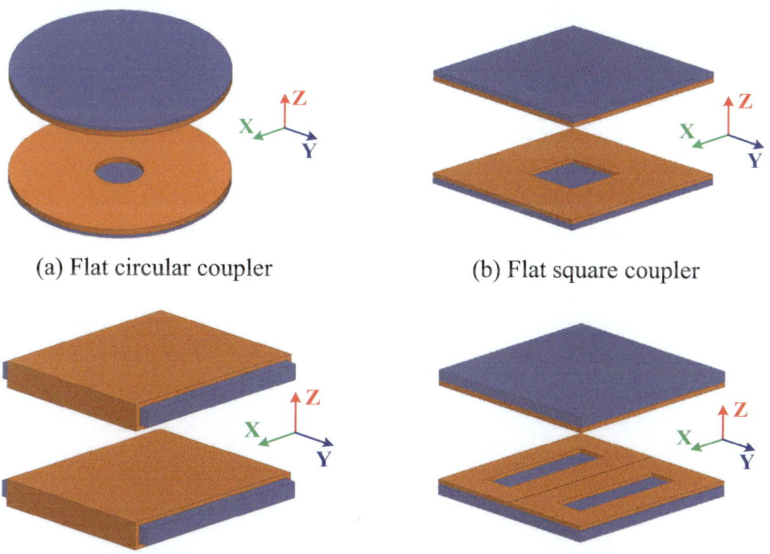

(a) Flat circular coupler (b) Flat square coupler

(c) Planar solenoid coupler (d) DD shaped coupler

Fig. 1.2 Four typical magnetic coupling structures in plane shape the misalignment tolerance capability of this structure. Pearce et al. [27] analyzed the DD-shaped magnetic coupling structure without magnetic core, improved the mechanical strength of this structure, and conducted experimental verification under high-power conditions. The WPT system based on the DD-shaped magnetic coupling structure was analyzed by Mohammad et al. [28] for loss, and the optimal method of magnetic core geometry design was analyzed and effectively limited the loss level of the magnetic coupling structure. Luo et al. [29] conducted a multi-objective optimization of the DD-shaped magnetic coupling structure through a non-dominated sorting genetic algorithm, reduced the magnetic leakage of the system under the premise of retainer coupling, and evaluated its anti-offset capability. However, due to the obvious difference between the *X*-axis and *Y*-axis directions of this structure, it only has strong resistance to offset in a single direction.

In addition to the above-mentioned basic magnetic coupling structure, the research team of the University of Auckland in New Zealand proposed a DDQ-shaped magnetic coupling structure based on the DD-shaped coil and analyzed its better misalignment tolerance capability [30]. In the application scenario of electric vehicles, [31] analyzed a road power supply system based on DDQ magnetic structure and conducted analysis and research on its energy transmission, endurance, thermal balance, economy, etc. The feasibility of using asymmetric Archimedean coils to achieve strong misalignment tolerance capability of magnetic coupling structures was analyzed by Aditya et al. [32]. A planar magnetic coupling structure with a "Tai Chi" shape was proposed in [33], which has better anti-offset capability compared to planar circular and DD-shaped coils. A planar magnetic coupling structure with a triple-stack shape was proposed in [34], which improves its misalignment tolerance capability by effectively utilizing the spatial magnetic field distribution. The above-mentioned magnetic coupling structures further optimize the anti-offset capability of the WPT system. However, the relative air gap between the primary and secondary coils in the system is narrow, and the extension of the air gap along the *Z*-axis direction is less considered.

In summary, most of the existing research on planar magnetic coupling structures focuses on the analysis and research of their anti-migration capabilities and related optimization of coupling level, parameter design, weight, economy, and other indicators. In addition, conventional planar magnetic coupling structures are often targeted at WPT application scenarios with relatively narrow air gaps. They usually have larger coupling coefficients and higher energy transfer efficiency. However, when the distance between the primary and secondary edges of the magnetic structure increases, its coupling coefficient decreases significantly.

1.2.2 Three-Dimensional Shape Magnetically Coupled Structure

In order to expand the application scenarios of WPT technology and improve the comprehensive performance of WPT systems, relevant scholars have done a lot of work in improving the degree of freedom of WPT and extending the distance of WPT. Among them, it includes in-depth exploration of three-dimensional magnetic coupling structures.

In order to increase the dimensionality of WPT, the solution of using an orthogonal three-dimensional magnetic coupling structure to achieve omnidirectional WPT in space was proposed by the team of Professor Xu Shuyuan of the University of Hong Kong in [35, 36]. This solution modulates the vector direction of the magnetic field through a current control scheme to achieve omnidirectional WPT. A crossed magnetic dipole-shaped coupling mechanism suitable for wide-range omnidirectional WPT systems was proposed by the team of Professor Chun T. Rim of the Korea Institute of Science and Technology in [37]. This mechanism achieves WPT in any direction within a range of 1m. In addition, the modular orthogonal solenoid matrix magnetic coupling structure was proposed by this research group in [38], realizing wide-area WPT Wi-power similar to the role of WiFi in the communication engineering discipline. A bowl-shaped three-dimensional magnetic coupling structure was proposed by Professor Li Zeyuan's team at Virginia Tech in [39] to provide omnidirectional wireless charging for consumer electronics. A combined orthogonal omnidirectional WPT receiving coil was proposed by the research team of Tianjin University in [40, 41], achieving angle-insensitive omnidirectional WPT. In addition, an omnidirectional WPT system based on a three-coil cube-shaped magnetic coupling structure was proposed by scholars from Central South University in [42] and realized real-time omnidirectional wireless charging of multiple devices through a weighted algorithm. The degree of freedom of energy transmission in the three-dimensional space of the WPT system has been expanded by the above research, which relies on the directional divergence of the magnetic field to be controlled through relevant algorithms. However, due to the high dispersion of its space vector magnetic field, it does not have the conditions for magnetic field lines to converge, making it difficult to achieve WPT with long air gaps.

In order to improve the misalignment tolerance capability of WPT systems, scholars have proposed corresponding three-dimensional shape magnetic coupling structures with strong misalignment tolerance capability through research on three-dimensional magnetic coupling structures. In [43], the researchers proposed an asymmetrically shaped magnetic coupling structure based on the magnetic field concentration theory and achieved strong misalignment tolerance capability of the magnetic structure through optimized design methods. A Korean research team proposed an asymmetric π-shaped magnetic coupling structure for static charging of electric vehicles in [44] and proposed a complex vector flux simulation suitable for arbitrary resonant coils in inductive WPT systems. Advantage field analysis method. And the team modeled and analyzed it through the hysteresis loss model. A hybrid solenoid magnetic coupling structure was proposed by researchers from the University of Auckland in [45]. Through the three-dimensional composite improvement of the DD-shaped magnetic structure, the new magnetic coupling structure has both strong anti-offset capability and level of gap WPT with wide airflow. The above research conducted three-dimensional optimization based on conventional planar magnetic coupling structures, which further enhanced the misalignment tolerance capability of the magnetic coupling structure, but the improvement in WPT capabilities under long air gaps was limited.

In order to extend the distance of WPT and maintain a certain energy transmission efficiency, relevant scholars have conducted research on magnetic coupling structures with wide air gap WPT capabilities. Among them, a square solenoid coupler was proposed by scholars in [46], as shown in Fig. 1.3a. Through the analysis of influencing factors such as skin effect and proximity effect, measures to improve the quality factor are proposed, and the corresponding frequency corresponding to the optimal quality factor is obtained through theoretical analysis, achieving high wireless energy transmission efficiency under a wide air gap and low coupling coefficient situation. A three-dimensional hollow solenoid magnetic coupler was proposed by a related research team in [47], as shown in Fig. 1.3b. This coupler achieves 1.5 m WPT through a coil with a diameter of 40 cm. A new type of dual intermediate resonant circuit WPT system was proposed by researchers in [48], which makes the distance between the receiving coil and the transmitting coil greater than four times the diameter of the resonant coil. Some scholars improve the quality factor of the magnetic coupling structure by increasing the operating frequency of the system to further achieve wide air gap WPT. The current and magnetic fields of current-carrying coils in the system were analyzed by Lee and Lorenz [49], showing that the coils can be represented by very weak mutual inductances and as two collective inductors and resistors. All inductances and resistances are estimated through closed form expressions. Based on the proposed model and estimated circuit parameters, the power transfer efficiency equation was derived, and high-efficiency WPT was achieved at a frequency of 3.7 MHz and an air gap of 30 cm. A cylindrical magnetic coupling structure for high-frequency WPT was proposed in [50] and achieved about 70% strong anti-offset WOT at 27.12 MHz. The three-dimensional copper ring magnetic coupling structure was obtained by advanced 3D casting method in [51], achieving wide air gap WPT at 6.78 and 13.56 MHz. The above research has optimized the design of the coil, based on related methods of reducing resistance or increasing frequency, and in the process of extending the air gap of WPT, by improving the quality factor of the magnetic coupling structure to compensate for the reduction of its coupling coefficient, thereby achieving WPT with long air gap. However, in order to achieve meter-level WPT and minimize resistance, the size of the magnetic structure is often larger. In addition, the high cost of high-level low-resistance coils or megahertz-level power devices and their control modules reduces the economics of the system.

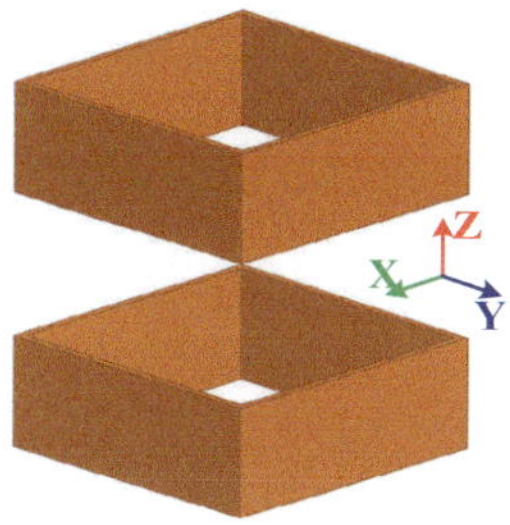

(a) Square solenoid coupler

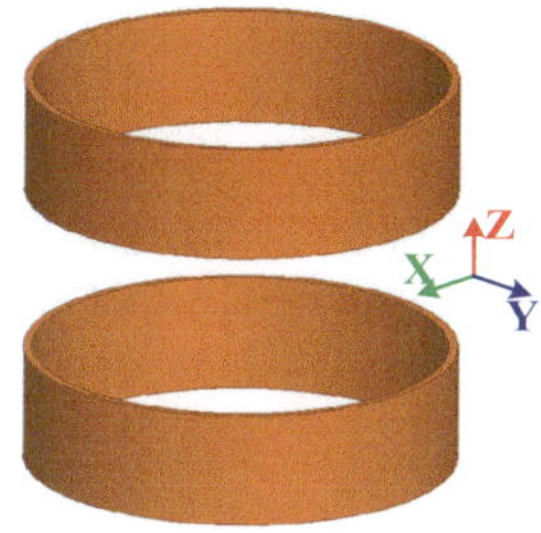

(b) Hollow solenoid coupler

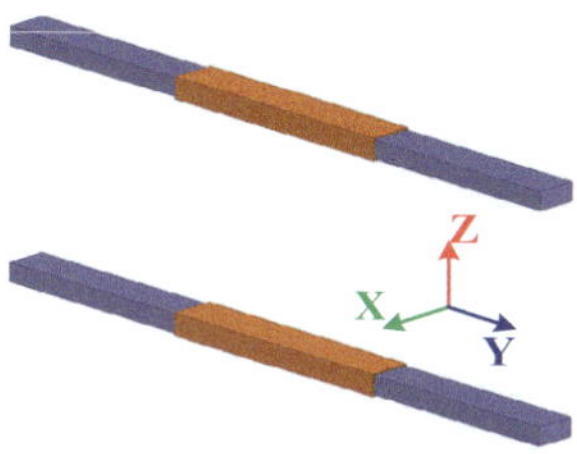

(c) Magnetic dipole coupler

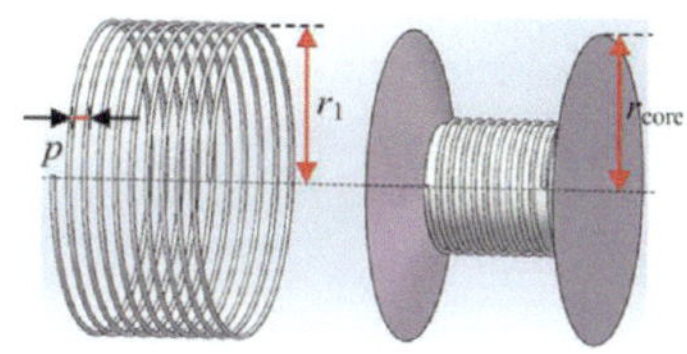

(d) Double funnel shaped coupler

Fig. 1.3 Four magnetic coupling structures in 3D shape

Nonlinear materials with high magnetic permeability such as ferrite cores are widely used in WPT systems. Due to their high magnetic permeability characteristics, they can effectively enhance the coupling degree between the primary and secondary magnetic coupling structures and play the role of improving the coupling coefficient. A three-dimensional magnetic coupling structure in the shape of a magnetic dipole was proposed by the team of Professor Chun T. Rim of the Korea Institute of Science and Technology in [9, 52–54]. This structure can be applied to wide air gap WPT, as shown in Fig. 1.3c. Core losses were analyzed deeply in [9], and a stepped core distribution method was proposed to make the magnetic field uniformly distributed in this structure. The use of variable switched capacitance was proposed in [52] as a method to achieve load power regulation when the distance between the transmitter and receiver coils varies greatly. The magnetizing inductance of the primary coil and the secondary coil were compared in [53], which verified the overall superiority of the magnetic dipole coupling structure in long-distance power transmission. An optimal ampere-turn design for a long-distance WPT system based on a magnetic dipole coupling structure was proposed in [54] to achieve maximum energy transmission efficiency. An improved coupling design method for a non-contact sliding WPT system was proposed by a research team at the University of Auckland in [55] to achieve an improvement in its coupling coefficient within the same physical size range based on three-dimensional shapes such as U-shaped, E-shaped, and tank-shaped. The demagnetization factors of cylindrical ferrite core and the proposed double funnel-shaped ferrite core were studied by scholars in [56]. Based on the analytical expression of mutual inductance, concluding that weakening the demagnetization factor of the ferrite coil can improve the magnetic field of the coil, and based on this, the WPT distance was extended, as shown in Fig. 1.3d. In addition, a WPT system based on the metamaterial magnetic coupling structure of cubic high dielectric resonators was proposed by scholars in [57]. This system has the characteristics of strong coupling, high efficiency, miniaturization, and long-range WPT and extends the distance of WPT by introducing new materials. The above research is based on high magnetic permeability materials and achieves long air gap WPT by improving the coupling coefficient of the magnetic coupling structure. However, the size of the relevant magnetic coupling structures is usually larger. Although the air gap length of WPT is extended, the relative distance of energy transmission is low and the system efficiency is low. Therefore, it is necessary to conduct a more in-depth exploration of this field based on the above research and provide a generalized design method for related magnetic coupling structures.

1.3 The Impact of Coil Parasitic Parameters

1.3.1 Coil Model

A coil is not a pure inductor; the conductor has losses, and there is parasitic capacitance between the turns of the coil. By using a coil model, its high-frequency characteristics and losses can be analyzed. Paper [58] proposed the coil equivalent model shown in Fig. 1.4, which mainly includes three parts: inductance, parasitic resistance, and parasitic capacitance. Based on this model, the quality factor of the coil can be approximated. However, the above model is only valid at low frequencies, and the value of the equivalent parasitic capacitance C_L under high frequency conditions is no longer constant. Therefore, some researchers established a model of the distribution parameters of the coil [59], as shown in Fig. 1.5. The coils are divided into multiple segments and the parasitic capacitance and mutual inductance between the coils are taken into account, which is suitable for circuit analysis and calculation at high frequencies. Based on this distributed parameter model, the high-frequency equivalent inductance of the coil can be calculated more accurately. However, the quality factor (Q-factor) of the coil cannot be precisely determined because the losses associated with the parasitic capacitance are not considered in the model.

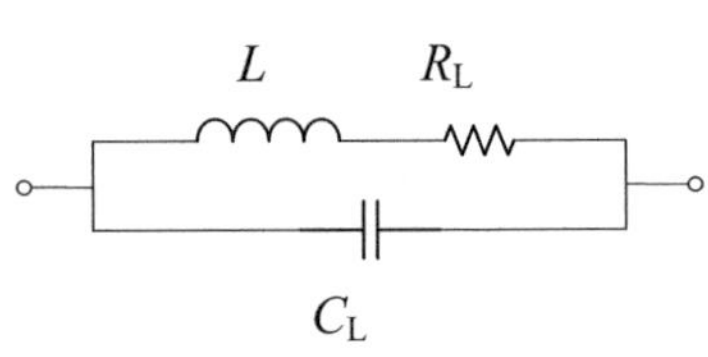

Fig. 1.4 Equivalent circuit model of the coil

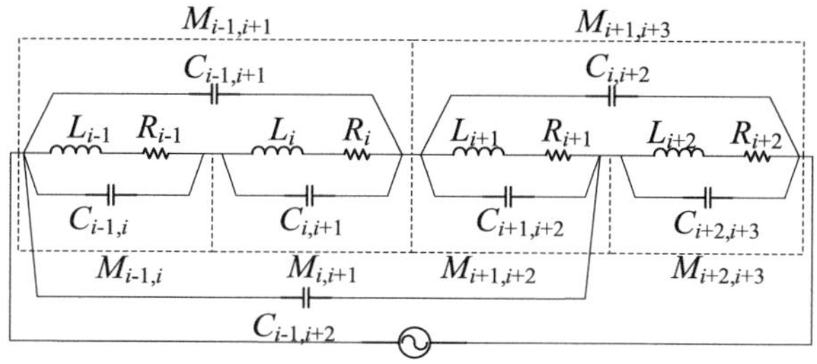

Fig. 1.5 Equivalent circuit model of coil distribution parameters

1.3.2 The Solution and Utilization of Coil Parasitic Capacitance Issues

Some researchers have studied the quality factor improvement technology of coils, and the parasitic capacitance has a great influence on the equivalent inductance and quality factor of LCT single-sided coils. The primary solution is to reduce the value of the parasitic capacitance. According to the model shown in Fig. 1.5, it is found that increasing the spacing of the coil can reduce the influence of parasitic capacitance, and the quality factor of the coil is improved by 10%. Paper [60] proposes a new basket coil as shown in Fig. 1.6a. By increasing the spacing between adjacent turns of the coils, the parasitic capacitance between the turns of the coils is effectively reduced, and its quality factor is improved. In order to avoid the adverse effects of coil parasitic capacitance, some researchers use single-layer solenoid coils for power transfer [46, 61–63], as shown in Fig. 1.6b. Single-layer solenoid coils have the characteristics of small parasitic capacitance and high coupling coefficient of long-distance transmission and are often used in applications with high transmission distance and diameter ratios. The above method to solve the adverse effects of parasitic capacitance is mainly to reduce the value of parasitic capacitance by increasing the coil spacing, thereby reducing its influence on the resonant circuit and reducing the system loss. However, the increase in turn spacing also increases the volume of the coil, and the increase in system power density is limited.

The role of parasitic capacitance is not entirely disadvantageous, and some researchers use the self-capacitance of coils and conductors to achieve power transfer in systems with only a single wire. 135-W power is transmitted at 18% efficiency at a transmission distance of 200 m [64], and the distributed parameter model of this single-wire power transfer system is shown in Fig. 1.7. According to this model, the frequency splitting phenomenon of the system is analyzed, and the resonant frequency of the system is accurately calculated. In addition, the parasitic capacitance of the coil is used for resonant wireless power transfer [7, 65, 66]. The small coil excites the large coil by inductive coupling, and the parasitic capacitance of the large coil is used to compensate for the inductive reactive power of the coil. A schematic diagram of the system circuit is shown in Fig. 1.8. Because the value of the parasitic capacitance is generally small, the resonant frequency of such systems

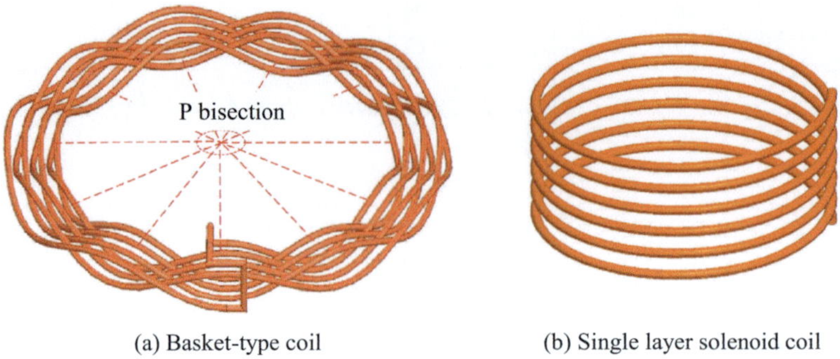

Fig. 1.6 Basket-type coil and single layer solenoid coil

is high, usually in the megahertz range. Therefore, it is not suitable for high-power applications.

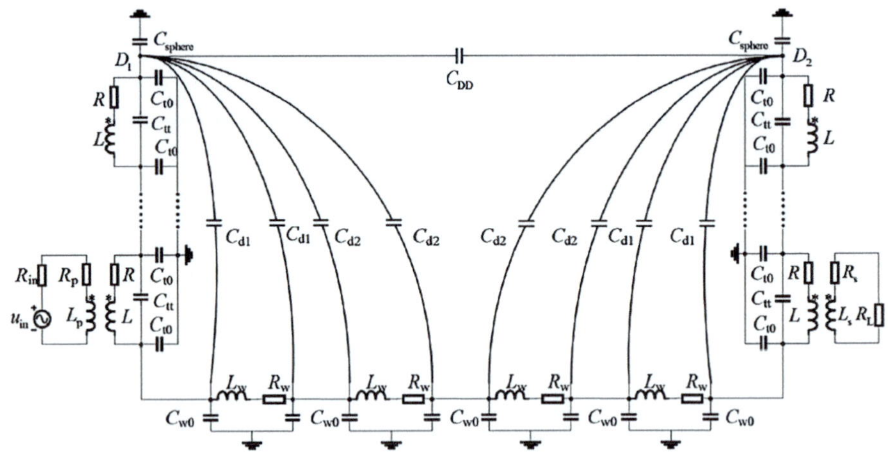

Fig. 1.7 Equivalent circuit of single-wire power transmission

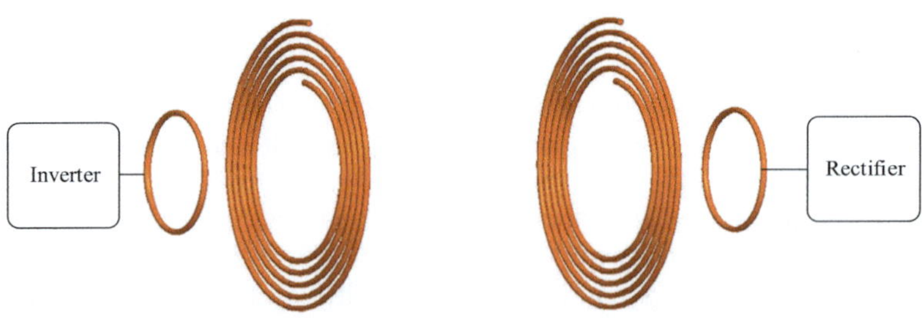

Fig. 1.8 Schematic diagram of a four-coil system

The effect of parasitic capacitance is inseparable from the voltage stress of the coil. Generally, the higher the coil voltage, the greater the voltage of the parasitic capacitance. In order to reduce the coil voltage stress, the researchers proposed a method of distributed compensation to reduce the voltage of the parasitic capacitance. The resonant capacitors are connected in series to the middle of the inductor coil to solve the problem of excessive coil voltage stress. Paper [67] proposes a method of segmented series capacitance for resonant inductors. The resonant capacitors are connected in series to the middle of the inductor coil to solve the problem of excessive coil voltage stress. Korea Railway Research Institute designed a 1 MW IPT system for high-speed trains. Due to the long transmission line and high power, the voltage stress of the system is very high. In order to reduce the voltage stress, the design team used distributed compensation [68], as shown in Fig. 1.9. In addition, Paper [69] designed an IPT system for implantable medical applications. The high voltage stress of the coil can pose a safety hazard for this technology. Therefore, the method of distributed compensation is adopted, and the electric field distribution of centralized compensation and segmented compensation is shown in Fig. 1.10. The left side is centralized compensation, the right side is distributed compensation, and the comparison shows that the strength of the electric field of the distributed compensation coil is much smaller than that of centralized compensation. However, the above segmented compensation method is only for the consideration of reducing the coil voltage stress and does not reduce the voltage of the parasitic capacitance, so the transmission efficiency of the loosely coupled transformer is not significantly improved.

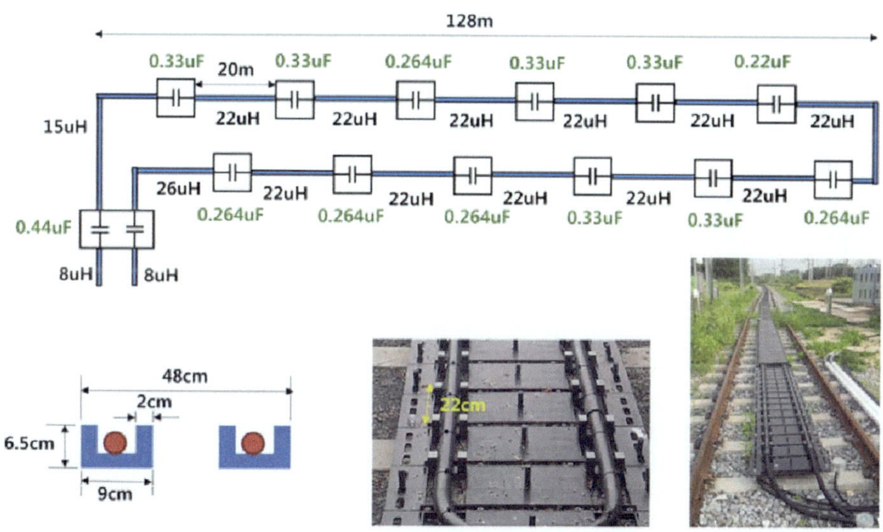

Fig. 1.9 The circuit structure diagram and physical diagram of the transmitter of the wireless power supply system for the high-speed train

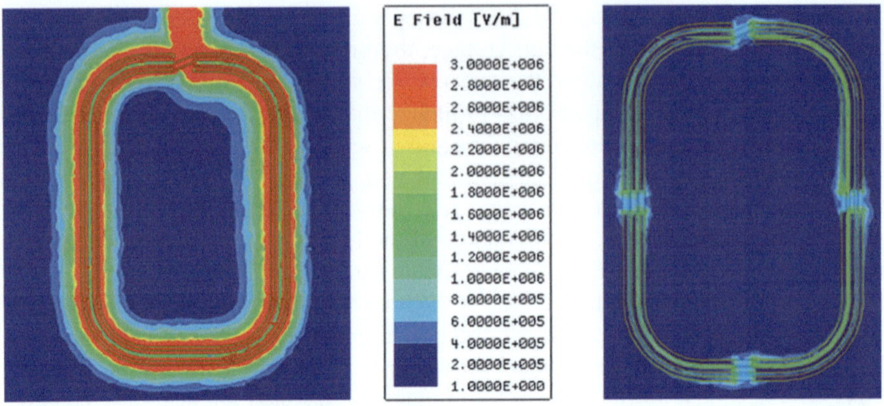

Fig. 1.10 Electric field distribution diagrams of coils with collective compensation and segment compensation

1.4 Rectifier Input Impedance Characteristics

Reactive power has a considerable impact on the efficiency of long air-gap wireless power transfer systems. The harmonics introduced by the rectifier circuit can influence the impedance characteristics of the load, potentially causing an increase in reactive power. Some studies approximate the equivalent AC impedance of the load as purely resistive. The calculation of the AC resistance R_{re} of the C-filtered and LC-filtered full-bridge rectifier circuit is shown in formula (1.1); R_L is the load resistance value. However, this approximate equivalence has a large error when the load resistance is large (C-filtered) or small (LC-filtered). The input voltage and current of the discontinuous conduction mode (DCM) rectifier are distorted, resulting in an increase in the phase difference between the input voltage and current of the rectifier [70]. The paper [71] designed a 300 kW high-power wireless power transfer system, and the inductive load-equivalent AC impedance (LEACI) not only increased the transmission of reactive power but also seriously affected the implementation of zero voltage switching (ZVS) of the inverter. Some researchers replace the LEACI of the C-filtered rectifier with an inductor series resistance [72–75]. The efficiency of systems with relatively long wireless power transfer is very sensitive to the reactive power. The inductive LEACI causes the IPT system to transmit reactive power, so some researchers try to solve this problem by adjusting the secondary side compensation parameters to achieve the series resonance of the compensation capacitance, coil, and the inductance component of LEACI [72]. The main limitation of this method is that the inductive component of the equivalent AC load is not fixed and varies greatly with the load.

$$\begin{cases} R_{re} = \frac{8}{\pi^2} R_L \text{ (C filtering)} \\ R_{re} = \frac{\pi^2}{8} R_L \text{ (LC filtering)} \end{cases} \tag{1.1}$$

Professor John T. Boys of the University of Auckland analyzed the distortion of the current waveform of the heavy-duty rectifier in the LCL compensation topology and calculated the ratio of the rectifier output voltage to the secondary capacitor voltage in parallel under critical mode conditions according to the diode turn-on condition [76]. Professor Chris Mi's team of the University of Michigan proposed a method for calculating the load equivalent AC impedance angle (LEACIA) the under continuous conduction mode (CCM) through harmonic analysis. Based on the analysis of the AC impedance of the load, a method is proposed to eliminate the influence of the inductive LEACI on the pre-circuit by changing the series capacitance value of the secondary side of the LCC/LCC compensation topology, which not only reduces the transferred reactive power of transformer but also reduces the fluctuation of output current [77, 78]. Based on the volt-second balance of the inductor, the team also proposed a calculation method of the LEACI for the secondary-side LCC compensation topology under continuous mode and intermittent mode. The secondary-side LCC compensation topology is shown in Fig. 1.11 [79]. Because the analysis model of DCM is relatively simplified, there is a small error in the calculation results. Some researchers have used the same method to study the rectifier loading characteristics of LCC compensation topology and improved the system model accordingly. The state of charge of the load is identified by the primary side voltage measurement, and the error is reduced compared to the identification of the equivalent pure resistive load [80, 81]. Paper [82] analyzes the nonlinear characteristics of rectifier loads, and the results show that the inductive characteristics of rectifier loads cannot be ignored. According to the inductive characteristics of rectified loads, an optimal design method of T-shaped matching network is proposed. Compared to traditional methods, the efficiency of the system is increased by nearly 1%. The team also analyzed the continuous and discontinuous modes of the rectifier based on the LCC/LCC compensation topology, calculated the equivalent AC impedance of the load, and adjusted the compensation parameters accordingly [83]. Paper [84] increases the load range of CCM mode by adding an LC resonant filter circuit. The larger the inductance value, the wider the range of the load. But this approach sacrifices the size and cost of the system.

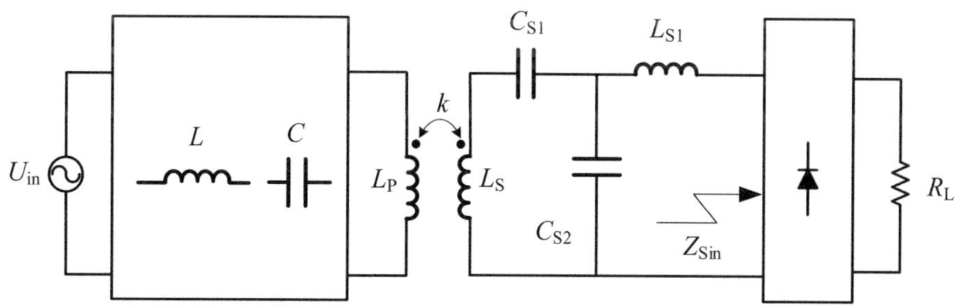

Fig. 1.11 Secondary LCC compensation topology

The above analysis is based on the C-filter rectifier. Less research has been done on the equivalent AC impedance characteristics of LC filter rectifier. Paper [85] analyzes the main reasons for the capacitive performance of the LEACI of the LC filter rectifier and calculates the current time domain waveform under continuous mode of S/SP compensation topology. Because the harmonic current flowing through the secondary coil is ignored, the calculation results are different from the experimental results. Paper [86] proposes an accurate calculation method for LEACI considering the harmonic currents of the primary and secondary coils based on the above research. The fundamental wave and each harmonic component of the rectifier input are solved by iterative method. The calculated current waveform of the S/SP compensation topology is more accurate and more consistent with the experimental waveform.

Paper [87] proposes a harmonic analysis method to analyze the main reasons why rectifiers enter DCM. For C filtering, when the load resistance is large, the compensation network cannot provide enough harmonic current to turn the diode on, and the rectifier enters DCM. For LC filtering, when the load resistance is small, the compensation network cannot provide enough harmonic voltage to turn the diode on, and the rectifier enters DCM. The team calculated the load boundary between CCM and DCM as a design criterion in detail and proposed an S/CLC compensation topology with the characteristics of high critical load resistance.

References

1. Zhang Z, Pang H, Georgiadis A et al (2019) Wireless power transfer—an overview. IEEE Trans Industr Electron 66(2):1044–1058
2. Fan X, Mo X, Zhang X (2015) Research status and application of wireless power transmission technology. Proc CSEE 35(10):2584–2600 (in Chinese)
3. Cheng S, Chen X, Wang J, Wen J, Li J (2015) Key technologies and applications of wireless power transmission. Trans China Electrotech Soc 30(19):68–84 (in Chinese)
4. Marincic AS (1982) Nikola Tesla and the wireless transmission of energy. IEEE Trans Power Appar Syst PAS-101(10):4064–4068
5. Budhia M, Covic GA, Boys JT (2011) Design and optimization of circular magnetic structures for lumped inductive power transfer systems. IEEE Trans Power Electron 26(11):3096–3108
6. Covic GA, Boys JT (2013) Modern trends in inductive power transfer for transportation applications. IEEE J Emerg Sel Top Power Electron 1(1):28–41
7. Kurs A, Karalis A, Moffatt R et al (2007) Wireless power transfer via strongly coupled magnetic resonances. Science 317(5834):83–86
8. Zhang B, Shu X, Wu L, Rong C (2019) Problems of wireless power transmission technology urgent to be solved and corresponding countermeasures. Autom Electr Power Syst 43(18):1–12 (in Chinese)
9. Park C, Lee S, Cho GH et al (2015) Innovative 5-m-off-distance inductive power transfer systems with optimally shaped dipole coils. IEEE Trans Power Electron 30(2):817–827
10. Li Z, Lu R, Zhu C, Song K, Wei G (2015) Study on the performance optimization of long range and low power wireless power transmission system. Trans China Electrotech Soc 30(S1):209–214 (in Chinese)
11. Zhong W, Lee CK, Hui S (2013) General analysis on the use of Tesla's resonators in domino forms for wireless power transfer. IEEE Trans Industr Electron 60(1):261–270

12. Huang Z, Wang L, Zou J (2017) The influence of coil location parameters to load power in wireless power transmission with two or three relay coils. Trans China Electrotech Soc 32(05):208–214 (in Chinese)
13. Li Y, Liu G, Zhang C, An H (2016) Broadband power converter for wireless power transfer system via coupled magnetic resonances. Trans China Electrotech Soc 31(S1):25–31 (in Chinese)
14. Belo D, Ribeiro DC, Pinho P et al (2019) A selective, tracking, and power adaptive far-field wireless power transfer system. IEEE Trans Microw Theory Tech 67(9):3856–3866
15. Zhao S, Zhao H, Shou P (2020) Discussion on key technology and operation & maintenance of intelligent power equipment. Autom Electr Power Syst 44(20):1–10 (in Chinese)
16. Xiong L, Song D, He Y, Lin Y, Xu M, He W (2012) Power supply of online monitoring equipment for transmission line. Electr Power Autom Equip 32(07):129–133+139 (in Chinese)
17. Huang M, Tang N, Sheng C, Lu Q, Zeng J (2020) Composite insulator energy transmission structure for power supply of intelligent monitoring instruments. High Volt Eng 46(02):618–625 (in Chinese)
18. Acero J, Carretero C, Lope I et al (2013) Analysis of the mutual inductance of planar-lumped inductive power transfer systems. IEEE Trans Industr Electron 60(1):410–420
19. Yilmaz T, Hasan N, Zane R et al (2017) Multi-objective optimization of circular magnetic couplers for wireless power transfer applications. IEEE Trans Magn 53(8):1–12
20. Sallán J, Villa JL, Llombart A et al (2009) Optimal design of icpt systems applied to electric vehicle battery charge. IEEE Trans Ind Electron 56(6):2140–2149
21. Zhu Q, Zhang Y, Liao C et al (2017) Experimental study on asymmetric wireless power transfer system for electric vehicle considering ferrous chassis. IEEE Trans Transp Electrif 3(2):427–433
22. Luo Z, Wei X (2017) Analysis of square and circular planar spiral coils in wireless power transfer system for electric vehicles. IEEE Trans Industr Electron 65(1):331–341
23. Li Z, Zhu C, Jiang J et al (2016) A 3-kW wireless power transfer system for sightseeing car supercapacitor charge. IEEE Trans Power Electron 32(5):3301–3316
24. Nagatsuka Y, Ehara N, Kaneko Y et al (2010) Compact contactless power transfer system for electric vehicles. In: The 2010 international power electronics conference-ECCE ASIA. IEEE, pp 807–813
25. Yao Y, Gao S, Wang Y et al (2020) Design and optimization of an electric vehicle wireless charging system using interleaved boost converter and flat solenoid coupler. IEEE Trans Power Electron 36(4):3894–3908
26. Budhia M, Boys JT, Covic GA et al (2011) Development of a single-sided flux magnetic coupler for electric vehicle IPT charging systems. IEEE Trans Industr Electron 60(1):318–328
27. Pearce MGS, Covic GA, Boys JT (2018) Robust ferrite-less double D topology for roadway IPT applications. IEEE Trans Power Electron 34(7):6062–6075
28. Mohammad M, Choi S, Elbuluk ME (2019) Loss minimization design of ferrite core in a DD-coil-based high-power wireless charging system for electrical vehicle application. IEEE Trans Transp Electrif 5(4):957–967
29. Luo Z, Wei X, Pearce MGS et al (2021) Multiobjective optimization of inductive power transfer double-D pads for electric vehicles. IEEE Trans Power Electron 36(5):5135–5146
30. Covic GA, Kissin MLG, Kacprzak D et al (2011) A bipolar primary pad topology for EV stationary charging and highway power by inductive coupling. In: 2011 IEEE energy conversion congress and exposition. IEEE, pp 1832–1838
31. Zhang Z, Pang H, Lee CHT et al (2017) Comparative analysis and optimization of dynamic charging coils for roadway-powered electric vehicles. IEEE Trans Magn 53(11):1–6
32. Aditya K, Sood VK, Williamson SS (2017) Magnetic characterization of unsymmetrical coil pairs using Archimedean spirals for wider misalignment tolerance in IPT systems. IEEE Trans Transp Electrif 3(2):454–463
33. Li Y, Zhao J, Yang Q et al (2019) A novel coil with high misalignment tolerance for wireless power transfer. IEEE Trans Magn 55(6):1–4

34. Kim S, Covic GA, Boys JT (2016) Tripolar pad for inductive power transfer systems for EV charging. IEEE Trans Power Electron 32(7):5045–5057
35. Ng WM, Zhang C, Lin D et al (2014) Two-and three-dimensional omnidirectional wireless power transfer. IEEE Trans Power Electron 29(9):4470–4474
36. Zhang C, Lin D, Hui SY (2015) Basic control principles of omnidirectional wireless power transfer. IEEE Trans Power Electron 31(7):5215–5227
37. Choi BH, Lee ES, Sohn YH et al (2015) Six degrees of freedom mobile inductive power transfer by crossed dipole Tx and Rx coils. IEEE Trans Power Electron 31(4):3252–3272
38. Lee ES, Sohn YH, Choi BG et al (2018) A modularized IPT with magnetic shielding for a wide-range ubiquitous Wi-power zone. IEEE Trans Power Electron 33(11):9669–9690
39. Feng J, Li Q, Lee FC et al (2019) Transmitter coils design for free-positioning omnidirectional wireless power transfer system. IEEE Trans Industr Inf 15(8):4656–4664
40. Zhang Z, Zhang B (2019) Angular-misalignment insensitive omnidirectional wireless power transfer. IEEE Trans Industr Electron 67(4):2755–2764
41. Zhang Z, Zhang B, Wang J (2018) Optimal design of quadrature-shaped pickup for omnidirectional wireless power transfer. IEEE Trans Magn 54(11):1–5
42. Tang W, Zhu Q, Yang J et al (2019) Simultaneous 3-D wireless power transfer to multiple moving devices with different power demands. IEEE Trans Power Electron 35(5):4533–4546
43. Yao Y, Wang Y et al (2019) A novel unsymmetrical coupling structure based on concentrated magnetic flux for high-misalignment IPT applications. IEEE Trans Power Electron 34(4):3110–3123
44. Choi SY, Huh J, Lee WY et al (2014) Asymmetric coil sets for wireless stationary EV chargers with large lateral tolerance by dominant field analysis. IEEE Trans Power Electron 29(12):6406–6420
45. Tejeda A, Kim S, Lin FY et al (2018) A hybrid solenoid coupler for wireless charging applications. IEEE Trans Power Electron 34(6):5632–5645
46. Deng Q, Liu J, Czarkowski D et al (2016) Frequency-dependent resistance of litz-wire square solenoid coils and quality factor optimization for wireless power transfer. IEEE Trans Industr Electron 63(5):2825–2837
47. Cai C, Wang J, Liu R et al (2018) Resonant wireless charging system design for 110-kV high-voltage transmission line monitoring equipment. IEEE Trans Industr Electron 66(5):4118–4129
48. Liu X, Wang G (2015) A novel wireless power transfer system with double intermediate resonant coils. IEEE Trans Industr Electron 63(4):2174–2180
49. Lee SH, Lorenz RD (2011) Development and validation of model for 95%-efficiency 220-W wireless power transfer over a 30-cm air gap. IEEE Trans Ind Appl 47(6):2495–2504
50. Liu D, Georgakopoulos SV (2018) Cylindrical misalignment insensitive wireless power transfer systems. IEEE Trans Power Electron 33(11):9331–9343
51. Zulauf G, Rivas-Davila JM (2019) Single-turn air-core coils for high-frequency inductive wireless power transfer. IEEE Trans Power Electron 35(3):2917–2932
52. Lee ES, Choi BG, Choi JS et al (2016) Wide-range adaptive IPT using dipole-coils with a reflector by variable switched capacitance. IEEE Trans Power Electron 32(10):8054–8070
53. Choi BH, Thai VX, Lee ES et al (2016) Dipole-coil-based wide-range inductive power transfer systems for wireless sensors. IEEE Trans Industr Electron 63(5):3158–3167
54. Choi BG, Kim JH, Lee ES et al (2019) Optimal dipole-coil ampere-turns design for maximum power efficiency of IPT. IEEE Trans Power Electron 35(7):7317–7327
55. Abdolkhani A, Hu AP (2014) Improved coupling design of contactless slipring for rotary applications. IEEE J Emerg Sel Top Power Electron 3(1):288–295
56. Wang M, Feng J, Shi Y et al (2018) Demagnetization weakening and magnetic field concentration with ferrite core characterization for efficient wireless power transfer. IEEE Trans Industr Electron 66(3):1842–1851
57. Das R, Basir A, Yoo H (2018) A metamaterial-coupled wireless power transfer system based on cubic high-dielectric resonators. IEEE Trans Industr Electron 66(9):7397–7406
58. Ramrakhyani AK, Mirabbasi S, Mu C (2011) Design and optimization of resonance-based efficient wireless power delivery systems for biomedical implants. IEEE Trans Biomed Circuits Syst 5(1):48–63

59. Yang Z, Liu WT, Basham E (2007) Inductor modeling in wireless links for implantable electronics. IEEE Trans Magn 43(10):3851–3860
60. Li Y (2019) Research on high-frequency wireless power transfer technology with multiple receivers based on magnetic coupling resonance. Doctoral Dissertation, Harbin Institute of Technology (in Chinese)
61. Zhou H, Gao X, Lai J et al (2018) Natural frequency optimization of wireless power systems on power transmission lines. IEEE Access 6:14038–14047
62. Cai C, Wang J, Jing H et al (2019) Universal wireless powered terminals for robust overhead transmission line monitoring. IET Power Electron 12(14):3739–3748
63. Cai C, Wang J, Liu R et al (2019) Resonant wireless charging system design for 110-kV high-voltage transmission line monitoring equipment. IEEE Trans Industr Electron 6(5):4118–4129
64. Li T (2021) Research on single-wire electric field resonance wireless power transfer technology. Master's thesis, Dalian University of Technology (in Chinese)
65. Sample AP, Meyer DT, Smith JR (2011) Analysis, experimental results, and range adaptation of magnetically coupled resonators for wireless power transfer. IEEE Trans Industr Electron 58(2):544–554
66. Dang Z, Qahouq JAA (2014) Modeling and investigation of magnetic resonance coupled wireless power transfer system with lateral misalignment. In: 2014 IEEE applied power electronics conference and exposition—APEC 2014, pp 1317–1322
67. Chen Q, Yan K, Hou J, Ren X, Ruan X (2013) Resonant inductor segmented series capacitor resonant circuit. CN103296989A
68. Kim JH, Lee BS, Lee JH et al (2015) Development of 1-MW inductive power transfer system for a high-speed train. IEEE Trans Industr Electron 62(10):6242–6250
69. Zeng Y, Qiu D, Meng X, Zhang B, Tang SC (2018) Optimized design of coils for wireless power transfer in implanted medical devices. IEEE J Electromagn RF Microw Med Biol 2(4):277–285
70. Yao Y, Wang Y, Liu X et al (2018) A novel parameter tuning method for a double-sided LCL compensated WPT system with better comprehensive performance. IEEE Trans Power Electron 33(10):8525–8536
71. Wijaya FP, Shimotsu T, Saito T, Kondo K (2018) A simple active power control for a high-power wireless power transmission system considering coil misalignment and its design method. IEEE Trans Power Electron 33:9989–10002
72. Fu M, Tang Z, Liu M et al (2015) Full-bridge rectifier input reactance compensation in Megahertz wireless power transfer systems. In: 2015 IEEE PELS workshop on emerging technologies: wireless power (2015 WoW), pp 1–5
73. Berger A, Agostinelli M, Vesti S et al (2015) A wireless charging system applying phase-shift and amplitude control to maximize efficiency and extractable power. IEEE Trans Power Electron 30(11):6338–6348
74. Liu Y, Madawala UK, Mai R, He Z (2020) An optimal multivariable control strategy for inductive power transfer systems to improve efficiency. IEEE Trans Power Electron 35(9):8998–9010
75. Li J, Liao C, Wang L et al (2015) Research on decoupling design of maximum efficiency and transmission power for electric vehicle wireless charging system based on LCCL. Trans China Electrotech Soc 30(S1):199–203 (in Chinese)
76. Keeling NA, Covic GA, Boys JT (2010) A unity-power-factor IPT pickup for high-power applications. IEEE Trans Industr Electron 57(2):744–751
77. Li S, Li W, Deng J et al (2015) A double-sided LCC compensation network and its tuning method for wireless power transfer. IEEE Trans Veh Technol 64(6):2261–2273
78. Li W, Zhao H, Li S, Deng J et al (2015) Integrated LCC compensation topology for wireless charger in electric and plug-in electric vehicles. IEEE Trans Industr Electron 62(7):4215–4225
79. Zhang X, Kan T, You C, Mi C (2017) Modeling and analysis of AC output power factor for wireless chargers in electric vehicles. IEEE Trans Power Electron 32(2):1481–1492
80. Guo Y, Zhang Y, Zhang W, Wang L (2021) Battery parameter identification based on wireless power transfer system with rectifier load. IEEE Trans Industr Electron 68(8):6893–6904

81. Guo Y, Wang L, Zhang Y et al (2018) Rectifier load analysis for electric vehicle wireless charging system. IEEE Trans Industr Electron 65(9):6970–6982
82. Li S, Wang L, Guo Y et al (2017) Optimization of T-type impedance matching network design method for wireless charging systems based on rectifier load compensation. Trans China Electrotech Soc 32(24):9–16 (in Chinese)
83. Li S, Guo Y, Tao C et al (2019) Analysis of the input impedance of the rectifier and design of LCC compensation network of the dynamic wireless power transfer system. IET Power Electron 12(10):2678–2687
84. Li W, Zhang Q, Li H et al (2022) Series filter computational method for CCM recovery in double side LCC WPT system. IEEE Trans Industr Electron 69(9):8875–8882
85. Hou J, Chen Q, Ren X et al (2015) Time-domain characteristic analysis of S/SP contactless resonant converters. Proc CSEE 35(8):1983–1992 (in Chinese)
86. Chen Q, Fan F, Wang J et al (2022) Accurate calculation method of equivalent impedance for rectifier load based on S/SP compensation and optimization of inverter switching loss. 42(11):4138–4151 (in Chinese)
87. Qu X, Jing Y, Lian J et al (2019) Design for continuous-current-mode operation of inductive-power-transfer converters with load-independent output. IET Power Electron 12(10):2458–2465

Chapter 2
Compensation Topologies of Long Range IPT

Abstract In long range IPT system, the loose coupled transformer has a large leakage inductance, and the induced inductive reactive power is also large, which needs to be compensated with compensation elements. The compensation topology has an important impact on the transmission efficiency and output fluctuation of the system. Mainstream compensation topologies applicable to long range IPT systems mainly include S/S (primary series/secondary series), LCC/S (primary inductor—capacitor—inductor/secondary series), S/S with quad coils, LCC/LCC (primary inductor—capacitor—inductor/secondary inductor—capacitor—inductor), and S/S with multi-stage coils. This chapter describes the output characteristics of different compensation topologies and how impedance matching can be implemented to improve efficiency.

Keyword Compensation topology · Multi-stage coils · Impedance matching

2.1 S/S Compensation Topology

S/S compensation is the simplest and most common topology in IPT systems [1], as shown in Fig. 2.1. U_{in} is the output voltage of the inverter, U_{re} is the equivalent AC voltage of the load, L_P is the primary coil self-inductance, L_S is the secondary coil self-inductance, M is the mutual inductance, and R_{re} is the equivalent AC impedance of the load.

The compensation capacitors C_1 and C_2 resonate in series with L_P and L_S, respectively, as shown in formula (2.1), where ω_0 is the resonant frequency of the IPT system.

$$C_1 = \frac{1}{\omega_0^2 L_P}, C_2 = \frac{1}{\omega_0^2 L_S} \tag{2.1}$$

The input impedance Z_{in} and output current I_{re} satisfied formula (2.2) and (2.3). S/S compensation topology has the characteristics of load-independent constant current (CC) output and zero input impedance phase angle (ZPA).

Y. Wang et al., *Long Range Inductive Power Transfer*, CPSS Power Electronics Series,
https://doi.org/10.1007/978-981-96-7541-8_2

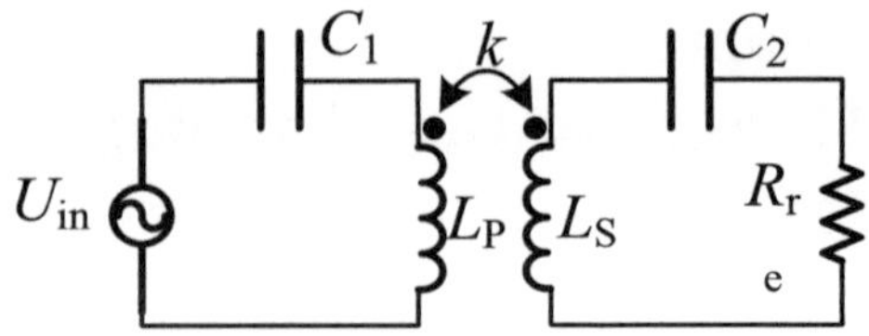

Fig. 2.1 S/S compensation topology

$$Z_{in} = \frac{(\omega_0 M)^2}{R_{re}} \tag{2.2}$$

$$\dot{I}_{re} = \frac{\dot{U}_{in}}{j\omega_0 M} \tag{2.3}$$

To make the system work at a higher level of transmission efficiency, the system needs to be impedance matched. Two ideas to realize impedance matching: If the design freedom of the secondary compensation circuit is high, impedance matching can be realized by changing the parameter size of the secondary compensation element. If the degree of freedom in the design of the secondary compensation circuit is low, the impedance matching can be realized by changing the number of windings or turns of the loose-coupled transformer, which in turn changes the relevant parameters of the loose-coupled transformer, such as the coupling coefficient k and the quality factor of the coil Q.

After considering the series parasitic resistors R_P and R_S of L_P and L_S, the efficiency of the system η is shown in formula (2.4).

$$\eta = \frac{M^2\omega_0^2 R_{re}}{(R_S + R_{re})(M^2\omega_0^2 + R_P R_S + R_P R_{re})} \tag{2.4}$$

From (2.4), the maximum efficiency load point R_{re_opt}, the maximum efficiency η_{max}, and the output power P_L are obtained, as shown in formula (2.5)–(2.7). Where Q_P and Q_S are the quality factors of L_P and L_S.

$$R_{re\_opt} = R_S\sqrt{1 + \frac{\omega_0^2 M^2}{R_S R_P}} = R_S\sqrt{k^2 Q_P Q_S + 1} \tag{2.5}$$

$$\eta_{max} = \frac{k^2 Q_P Q_S}{\left(\sqrt{k^2 Q_P Q_S + 1} + 1\right)^2} \tag{2.6}$$

$$P_L = \frac{{U_{in}}^2 R_S \sqrt{k^2 Q_P Q_S + 1}}{2\omega_0^2 M^2} \tag{2.7}$$

Considering the design freedom of the secondary side with the S/S compensation is not high, to achieve impedance matching, a change in the loosely coupled transformer is adopted. As shown in formula (2.4), adjusting k, R_S, Q_P, or Q_S can adjust R_{re_opt}.

2.2 LCC/S Compensation Topology

A common constant voltage output compensation topology is the LCC/S [2], as shown in Fig. 2.2.

L_1 is the compensation inductor, C_1–C_3 are compensation capacitors, which satisfy formula (2.8).

$$C_1 = \frac{1}{\omega_0^2 L_1}, C_2 = \frac{1}{\omega_0^2 (L_P - L_1)}, C_3 = \frac{1}{\omega_0^2 L_S} \tag{2.8}$$

The input impedance Z_{in} and output voltage U_{re} satisfied formula (2.9) and (2.10). LCC/S compensation topology also has the characteristics of load-independent constant voltage (CV) output and ZPA.

$$Z_{in} = \frac{L_1^2}{M^2} R_{re} \tag{2.9}$$

$$\dot{U}_{re} = \frac{M}{L_1} \dot{U}_{in} \tag{2.10}$$

The LCC/S compensation topology has a high degree of design freedom and the compensation inductor L_1 can be varied to adjust the output current.

After considering the series parasitic resistors R_P and R_S of L_P and L_S, the efficiency of the system η is shown in formula (2.11).

$$\eta = \frac{\omega_0^2 M^2 R_{re}}{(R_S + R_{re})^2 R_P + \omega_0^2 M^2 (R_S + R_{re})} \tag{2.11}$$

From (2.11), the maximum efficiency load point R_{re_opt}, and the output power P_L are obtained, as shown in formula (2.12) and (2.13).

$$R_{re\_opt} = R_S \sqrt{k^2 Q_P Q_S + 1} \tag{2.12}$$

$$P_L = \frac{M^2 U_{in}^2}{L_1^2 R_S \sqrt{k^2 Q_P Q_S + 1}} \tag{2.13}$$

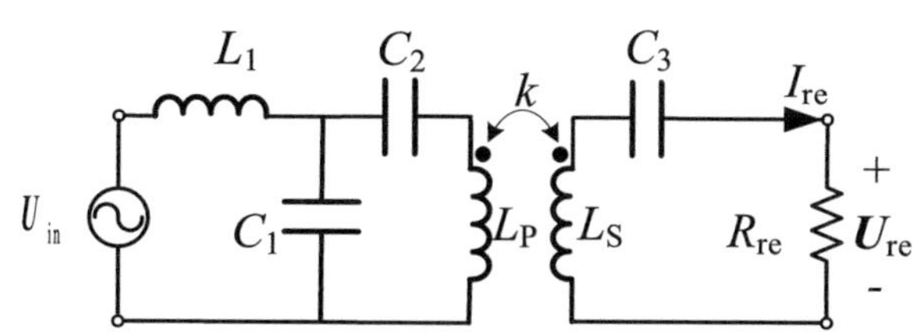

Fig. 2.2 LCC/S compensation topology

Considering the design freedom of the secondary side with the LCC/S compensation is not high, to achieve impedance matching, a change in the loosely coupled transformer is adopted. As shown in formula (2.12), adjusting k, R_S, Q_P, or Q_S can adjust R_{re_opt}.

2.3 LCC/LCC Compensation Topology

Due to design freedom limitations, higher order LCC/LCC compensation topologies are used [3], as shown in Fig. 2.3.

L_1, L_2 are compensation inductors and C_1–C_4 are compensation capacitors, which satisfy formula (2.14).

$$C_1 = \frac{1}{\omega_0^2 L_1},\ C_2 = \frac{1}{\omega_0^2 (L_P - L_1)},\ C_3 = \frac{1}{\omega_0^2 (L_S - L_2)},\ C_4 = \frac{1}{\omega_0^2 L_2} \tag{2.14}$$

$$Z_{in} = \frac{\omega_0^2 L_1^2 L_2^2}{M^2 R_{re}} \tag{2.15}$$

The LCC/LCC compensation topology has a high degree of design freedom and the compensation parameters L_1 and L_2 can be varied to adjust the output current.

$$\dot{I}_{re} = \frac{M \dot{U}_{in}}{j\omega_0 L_1 L_2} \tag{2.16}$$

After considering the series parasitic resistors R_P and R_S of L_P and L_S, the efficiency of the system η is shown in formula (2.17).

$$\eta = \frac{\omega_0^4 M^2 L_{f2}^2 R_{re}}{\left(R_P L_{f2}^4 + R_{re} L_{f2}^2 M^2\right)\omega_0^4 + \left(2L_{f2}^2 R_P R_S R_{re} + M^2 R_{re}^2 R_S\right)\omega_0^2 + R_{re}^2 R_S^2 R_P} \tag{2.17}$$

From (2.17), the maximum efficiency load point R_{re_opt} and the output power P_L are obtained, as shown in formula (2.18) and (2.19).

$$R_{re\_opt} = \frac{\omega_0^2 L_2^2}{R_S \sqrt{1 + \frac{\omega_0^2 M^2}{R_S R_P}}} = \frac{\omega_0^2 L_2^2}{R_S \sqrt{k^2 Q_P Q_S + 1}} \tag{2.18}$$

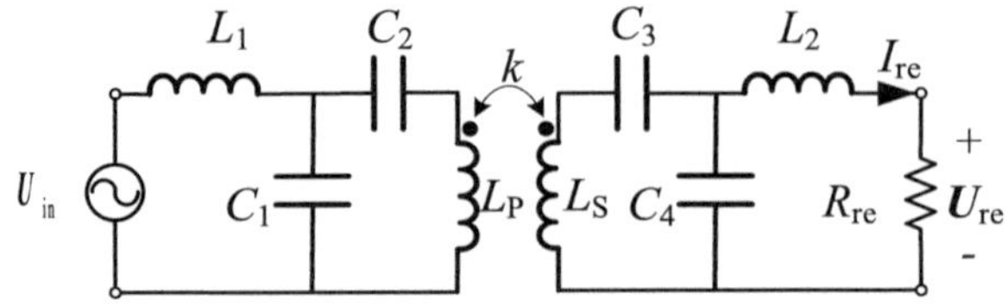

Fig. 2.3 LCC/LCC compensation topology

$$P_L = \frac{U_{in}^2 M^2}{2L_1^2 R_S \sqrt{k^2 Q_P Q_S + 1}} \tag{2.19}$$

Considering the design freedom of LCC/LCC compensation parameters is very high, to achieve impedance matching, a change in the compensation parameters is adopted. As shown in formula (2.18), adjusting L_2 can adjust R_{re_opt}.

2.4 Quad Coils with S/S Compensation Topology

To enhance the transmission distance of IPT system, quad coils is used [4], as shown in Fig. 2.4. The long-distance transmission of energy is realized through the relay coils L_2 and L_3, and the compensation topology often uses S/S compensation. When the resonant frequency between the coils is the same, the relay coil can effectively amplify the high-frequency current generated by the transmitting coil. L_1–L_4 are the self-inductances of the four coils in the magnetic coupler, C_1–C_4 are the series compensation capacitances of the corresponding coils, and M_{12}–M_{34} are the mutual inductance between each two coils.

In this four-coil structure, because the number of turns of the power coil and the load coil is small, the mutual inductance between them and the non-adjacent relay coils, and between them are very small and can be ignored. When the system is in full resonance, the coil series compensation capacitance satisfies the following equation:

$$C_1 = \frac{1}{\omega_0^2 L_1}, C_2 = \frac{1}{\omega_0^2 L_2}, C_3 = \frac{1}{\omega_0^2 L_3}, C_4 = \frac{1}{\omega_0^2 L_4} \tag{2.20}$$

The input impedance Z_{in} and output current I_4 satisfied formula (2.21) and (2.22). The quad coils with S/S compensation topology also have the characteristics of ZPA. Output current I_4 is related to R_{re} and therefore does not have load-independent constant voltage or constant current output characteristics.

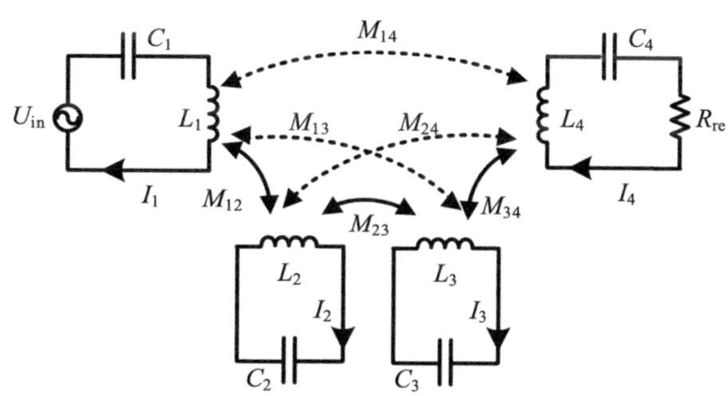

Fig. 2.4 Quad coils with S/S compensation topology

$$Z_{\text{in}} = \frac{\omega_0^2 M_{12}^2 M_{34}^2}{M_{23}^2 R_{\text{re}}} \tag{2.21}$$

$$\dot{I}_4 = -\frac{\text{j}\omega_0^3 M_{12} M_{23} M_{34} \dot{U}_{\text{in}}}{R_{\text{re}} + \omega_0^4 M_{12}^2 M_{34}^2} \tag{2.22}$$

After considering the series parasitic resistors R_1–R_4 of L_1 and L_4, the output power P_{L} and the efficiency of the system η are shown in formula (2.23) and (2.24).

$$P_{\text{L}} = \frac{\omega_0^6 M_{12}^2 M_{23}^2 M_{34}^2 U_{\text{in}}^2 R_{\text{re}}}{(R_1 R_2 + \omega_0^2 M_{12}^2) \cdot [R_3 (R_4 + R_{\text{re}}) + \omega_0^2 M_{34}^2] + \omega_0^2 M_{23}^2 R_1 (R_4 + R_{\text{re}})} \tag{2.23}$$

$$\eta = \frac{|\dot{I}_4|^2 R_{\text{re}}}{|\dot{I}_1|^2 R_1 + |\dot{I}_2|^2 R_2 + |\dot{I}_3|^2 R_3 + |\dot{I}_4|^2 (R_4 + R_{\text{re}})} \tag{2.24}$$

Under rated load, the parasitic resistance of the power supply coil L_1 and load coil L_4 is much smaller than the load and reflection resistance, so the efficiencies of L_1 and L_4 are close to 1. The AC efficiency depends on L_2 and L_3. Thus Eq. (2.25) can be simplified as:

$$\eta = \frac{\omega_0^2 M_{23}^2 R_{\text{eq4}}}{R_2 R_{\text{eq4}}^2 + (\omega_0^2 M_{23}^2 + 2R_2 R_3) R_{\text{eq4}} + \omega_0^2 M_{23}^2 R_3 + R_2 R_3^2}, \quad R_{\text{eq4}} = \frac{\omega^2 M_{34}^2}{R_4 + R_{\text{re}}} \tag{2.25}$$

The optimal load $R_{\text{re4\_opt}}$ corresponding to R_{re4} can be obtained by solving for the extreme value of η.

$$R_{\text{eq4}} = \frac{\omega_0^2 M_{34}^2}{R_4 + R_{\text{re}}} = R_3 \sqrt{\frac{\omega_0^2 M_{23}^2}{R_2 R_3} + 1} = R_3 \sqrt{1 + k_{23}^2 Q_{\text{L2}} Q_{\text{L3}}} \tag{2.26}$$

Optimal load matching for AC efficiency can be achieved by adjusting the coupling coefficient k_{34} of the receiver and power coils for impedance matching. Where Q_{L2} and Q_{L3} are the quality factors of L_2 and L_3.

2.5 Multi-stage Coils with S/S Compensation Topology

To extend the radio energy transmission distance, multi-stage coils are applied [5], which is a popularized form of quad coils, as shown in Fig. 2.5.

For ease of analysis, similar to quad coils, only the coupling of neighboring coils is usually considered.

When the system is in full resonance, the coil series compensation capacitance satisfies the following equation:

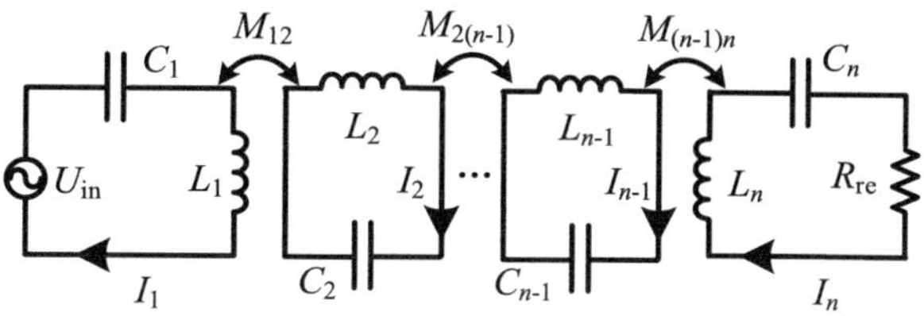

Fig. 2.5 Multi-stage coils with S/S compensation topology

$$C_1 = \frac{1}{\omega_0^2 L_1}, C_2 = \frac{1}{\omega_0^2 L_2}, \ldots, C_n = \frac{1}{\omega_0^2 L_n} \tag{2.27}$$

After considering the series parasitic resistors R_1–R_n of L_1 and L_n, the efficiency of the system η are shown in formula (2.28).

$$\eta = \frac{\left|\dot{I}_n\right|^2 R_{re}}{\left|\dot{I}_1\right|^2 R_1 + \left|\dot{I}_2\right|^2 R_2 + \cdots + \left|\dot{I}_{n-1}\right|^2 R_{n-1} + \left|\dot{I}_n\right|^2 (R_n + R_{re})} \tag{2.28}$$

The impedance matching analysis method is similar to the quad coils and will not be repeated here.

References

1. Wang CS, Stielau OH, Covic GA (2005) Design considerations for a contactless electric vehicle battery charger. IEEE Trans Indus Electron 52(5):1308–1314
2. Wang Y, Wang H, Liang T et al (2017) Analysis and design of an LCC/S compensated resonant converter for inductively coupled power transfer. In: 2017 IEEE transportation electrification conference and expo, Asia-Pacific (ITEC Asia-Pacific), pp 1–5
3. Liu C, Ge S, Guo Y et al (2016) Double-LCL resonant compensation network for electric vehicles wireless power transfer: experimental study and analysis. IET Power Electron 9(11):2262–2270
4. Moon S, Moon G-W (2016) Wireless power transfer system with an asymmetric four-coil resonator for electric vehicle battery chargers. IEEE Trans Power Electron 31(10):6844–6854
5. Liu Y, Chen J, Li Y, He Z (2024) A new modeling method for multiple-relay wireless power transfer system considering cross-coupling. IEEE Trans Industr Electron 71(2):1456–1467

Chapter 3
Strategies for Improving the Quality Factor of Coils

Abstract In a limited space, tightly wound coils are the main way to improve their quality factor. However, in IPT systems with relatively compact wound coil, parasitic capacitance can cause the coil to deviate from the pure inductance, destroy the resonant state of the compensation network, and increase system losses. There may be cases where the efficiency decreases when the amount of copper is increased. Previous research work has not paid attention to the dielectric loss caused by the low quality factor of parasitic capacitance. This chapter presents an accurate model of coil distribution parameters and reveals the mechanism of action of parasitic capacitance. Two solutions, single-turn segmented series compensation and single-layer segmented series compensation, are proposed to reduce the voltage stress of the parasitic capacitance and improve quality factor of coils and the transmission efficiency of LCT.

Keyword Factor of quality · Single-turn segmented series compensation · Single-layer segmented series compensation

In Chap. 2, the focus is on the compensation topologies for long-range inductive power transfer (IPT) systems, where the loosely coupled transformers result in significant leakage inductance and inductive reactive power, necessitating effective compensation strategies to enhance transmission efficiency and stability. However, as the transmission distance increases, the magnetic coupling weakens, posing greater challenges to energy transfer efficiency. Therefore, it becomes essential to improve transmission efficiency from another perspective, which is closely related to the quality factor of the coils [1]. This leads to the discussion in this chapter, which introduces an accurate distributed parameter model for coils, elucidates the mechanism of parasitic capacitance, and proposes solutions such as single-turn segmented series compensation and single-layer segmented series compensation to mitigate the voltage stress on parasitic capacitance, thereby improving both the quality factor and the transmission efficiency of loosely coupled transformers (LCT). This progression

Y. Wang et al., *Long Range Inductive Power Transfer*, CPSS Power Electronics Series,
https://doi.org/10.1007/978-981-96-7541-8_3

highlights the importance of addressing parasitic effects to achieve efficient long-range IPT systems, particularly in applications like long air-gap inductive wireless power transfer using the domino structure [2].

3.1 Model of the Single-Sided Coil of a Loosely Coupled Transformer

Under high-frequency conditions, the single-sided coil of the LCT can no longer be equivalent to pure inductance, and the equivalent inductance will change due to the parasitic capacitance of the coil. The quality factor of coils can be influenced by parasitic capacitance. In order to analyze this problem, it is necessary to establish a model of the distribution parameters of the coil, which not only needs to consider the copper loss of the coil but also cannot ignore the dielectric loss. According to the established coil distribution parameter model, the external characteristics and losses of the coil are analyzed.

3.1.1 Distributed Parameter Modeling of Coil

The coil is divided into n segments with their inductance values being L_1–L_n respectively, and the corresponding parasitic resistances being R_{L1}–R_{Ln}. There are parasitic capacitances and coupling relationships among various parts. $M_{(p,q)}$ represents the mutual inductance between the p-th and q-th segments, where $p = 1, 2, \ldots, n$, $q = 1, 2, \ldots, n$, and $p \neq q$. $C_{(p,q)}$ is the parasitic capacitance between the p-th and q-th segments, and its parasitic resistance is $R_{C(p,q)}$. Figure 3.1 shows the equivalent model I of the coil divided into n segments. The parasitic capacitances are equivalently connected to the mid-points of each coil segment. I_{L1}–I_{Ln} are the currents flowing into L_1–L_n respectively, U_{L1}–U_{Ln} are the mid-point voltages of inductors L_1–L_n, and $I_{C(p,q)}$ is the current of the parasitic capacitance $C_{(p,q)}$.

The parasitic capacitance connected to the middle of the coil is not convenient for calculating circuit parameters, and the equivalent model I of the coil is transformed to facilitate the calculation. L_m is divided into $L_{(m,1)}$ and $L_{(m,2)}$, R_{Lm} is divided into $R_{L(m,1)}$ and $R_{L(m,2)}$, $M_{[(m,1),(m,2)]}$ is the mutual inductance of $L_{(m,1)}$ and $L_{(m,2)}$, where $m = 1, 2, \ldots, n$, and they satisfy (3.1). $M_{[(p1,p2),(q1,q2)]}$ is the mutual inductance of $L_{(p1,p2)}$ and $L_{(q1,q2)}$, where $p_1 = 1, 2, \ldots n$, $q_1 = 1, 2, \ldots n$, $p_1 \neq q_1$, $p_2 = 1$, or 2, $q_2 = 1$ or 2, and they satisfy formula (3.2). The distribution parameter model II of the coil is shown in Fig. 3.2.

$$\begin{cases} L_{(m,1)}\big|_{m=1,2,\ldots,n} = L_{(m,2)} = \frac{1}{4}L_m \\ M_{[(m,1),(m,2)]}\big|_{m=1,2,\ldots,n} = \frac{1}{4}L_m \\ R_{L(m,1)}\big|_{m=1,2,\ldots,n} = R_{L(m,2)} = \frac{1}{2}R_m \end{cases} \tag{3.1}$$

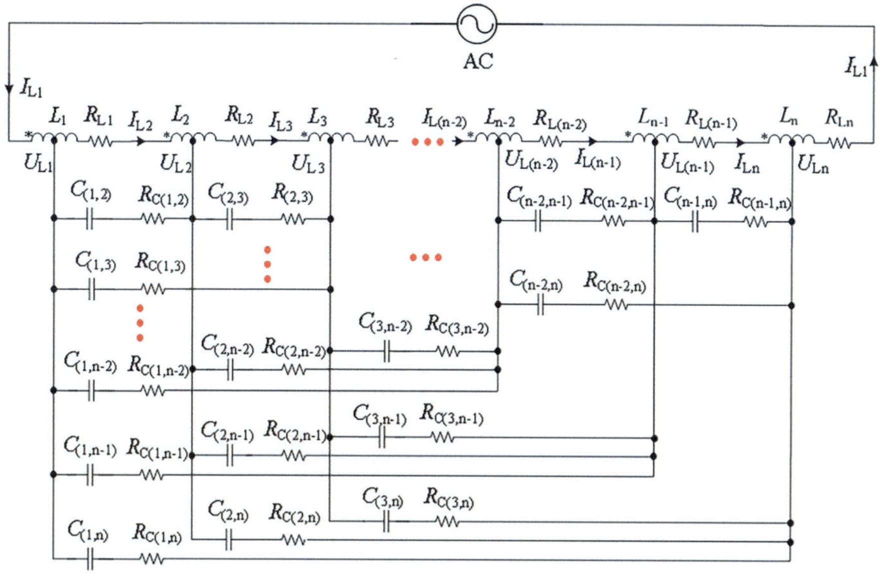

Fig. 3.1 Model I of the coil distribution parameters

$$M_{[(p1,p2),(q1,q2)]}\Big|_{(p1=1,2,\ldots,n)(q1=1,2,\ldots,n)(p1\neq q1)(p2=1,2)(q2=1,2)} = \frac{1}{4}M_{(p1,q1)} \tag{3.2}$$

By combining the inductance and resistance in series between the two nodes, the distribution parameter model II of the coil shown in Fig. 3.2 can be replaced by the distribution parameter model III shown in Fig. 3.3, and the variable parameters in the distribution parameter model III can be calculated by formula (3.3).

$$\begin{cases} L'_1 = L_{(1,1)} + L_{(n,2)} + 2M_{[(1,1),(n,2)]} \\ L'_m\big|_{m=2,3,\ldots,n} = L_{(m-1,2)} + L_{(m,1)} + 2M_{[(m-1,1),(m,2)]} \\ M'_{(1,q)}\Big|_{q=2,3,\ldots,n} = M_{[(n,2),(q-1,2)]} + M_{[(n,2),(q,1)]} + M_{[(1,1),(q-1,2)]} + M_{[(1,1),(q,1)]} \\ M'_{(p,q)}\Big|_{(p=2,3,\ldots,n)(q=2,3,\ldots,n)(p\neq q)} = M_{[(p-1,2),(q-1,2)]} + M_{[(p-1,2),(q,1)]} \\ \qquad + M_{[(p,1),(q-1,2)]} + M_{[(p,1),(q,1)]} \\ R'_1 = R_{(1,1)} + R_{(n,2)} \\ R'_{Lm}\big|_{m=2,\ldots,n} = R_{(m-1,2)} + R_{(m,1)} \end{cases} \tag{3.3}$$

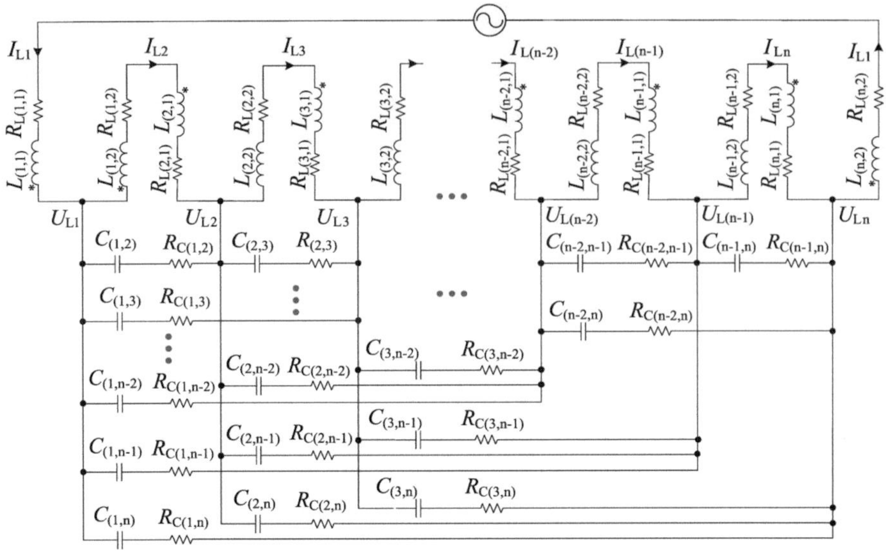

Fig. 3.2 Model II of the coil distribution parameters

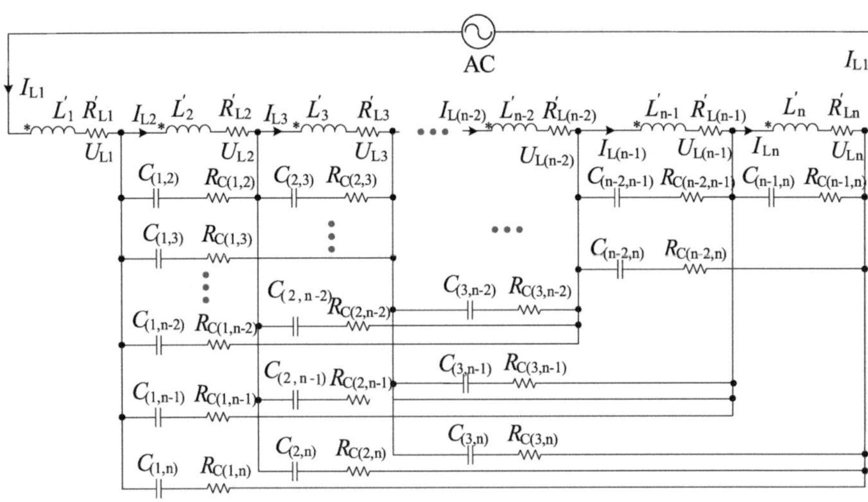

Fig. 3.3 Model III of the coil distribution parameters

3.1.2 Analysis of the External Characteristics and Loss of the Coil Based on the Distributed Parameter Model

The voltage and current parameters of the equivalent model III can be solved by Kirchhoff's voltage and current law. The parametric equation is as shown in (3.4), the matrix and vector in (3.4) is shown in formula (3.5)–(3.11) where A and B are parameter matrices composed of 1 and –1, "–" represents the flow node, U_{L} is the node voltage vector, I_{L} is the current vector of the inductor branch, Z_{L} is the impedance matrix composed of the branch inductance and its mutual inductance, Y_{C} is the admittance matrix of the capacitor branch, and U_{S} is the applied voltage source matrix. There are two ways to apply voltage, one is to directly connect the voltage source in series into the coil, and the other is to excite the coil by inductive coupling.

$$\begin{cases} AU_{\mathrm{L}} = Z_{\mathrm{L}}I_{\mathrm{L}} + U_{\mathrm{S}} \\ Y_{\mathrm{C}}U_{\mathrm{L}}+BI_{\mathrm{L}} = 0 \end{cases} \tag{3.4}$$

$$A = \begin{bmatrix} -1 & 0 & \ldots & 0 & 1 \\ 1 & -1 & \ldots & 0 & 0 \\ 0 & 1 & \ldots & 0 & 0 \\ \ldots & \ldots & \ldots & \ldots & \ldots \\ 0 & 0 & \ldots & 1 & -1 \end{bmatrix} \tag{3.5}$$

$$U_{\mathrm{L}} = \begin{bmatrix} U_1 \\ U_2 \\ \ldots \\ U_{\mathrm{n}} \end{bmatrix} \tag{3.6}$$

$$Z_{\mathrm{L}} = \begin{bmatrix} \mathrm{j}\omega L'_1 + R'_{L1} & \mathrm{j}\omega M'_{(1,2)} & \ldots & \mathrm{j}\omega M'_{(1,\mathrm{n})} \\ \mathrm{j}\omega M'_{(2,1)} & \mathrm{j}\omega L'_2 + R'_{L2} & \ldots & \mathrm{j}\omega M'_{(2,\mathrm{n})} \\ \ldots & \ldots & \ldots & \ldots \\ \mathrm{j}\omega M'_{(\mathrm{n},1)} & \mathrm{j}\omega M'_{(\mathrm{n},2)} & \ldots & \mathrm{j}\omega L'_{\mathrm{n}} + R'_{\mathrm{Ln}} \end{bmatrix} \tag{3.7}$$

$$I_{\mathrm{L}} = \begin{bmatrix} I_1 \\ I_2 \\ \ldots \\ I_{\mathrm{n}} \end{bmatrix} \tag{3.8}$$

$$U_{\mathrm{S}} = \begin{bmatrix} U_{\mathrm{S}1} \\ U_{\mathrm{S}2} \\ \ldots \\ U_{\mathrm{Sn}} \end{bmatrix} \tag{3.9}$$

$$Y_C = \begin{bmatrix} \sum_{q=1}^{n,q\neq p=1} -Y_{C(p,q)} & \frac{-j\omega C_{(1,2)}}{1+j\omega C_{(1,2)}R_{(1,2)}} & \cdots & \frac{-j\omega C_{(1,n)}}{1+j\omega C_{(1,n)}R_{(1,n)}} \\ \frac{-j\omega C_{(2,1)}}{1+j\omega C_{(2,1)}R_{(2,1)}} & \sum_{q=1}^{n,q\neq p=2} -Y_{C(p,q)} & \cdots & \frac{-j\omega C_{(2,n)}}{1+j\omega C_{(2,n)}R_{(2,n)}} \\ \cdots & \cdots & \cdots & \cdots \\ \frac{-j\omega C_{(n,1)}}{1+j\omega C_{(n,1)}R_{(n,1)}} & \frac{-j\omega C_{(n,2)}}{1+j\omega C_{(n,2)}R_{(n,2)}} & \cdots & \sum_{q=1}^{n,q\neq p=n} -Y_{C(p,q)} \end{bmatrix} \tag{3.10}$$

$$B = \begin{bmatrix} -1\ 1 & 0 & \ldots 0 \\ 0 & -1\ 1 & \ldots 0 \\ \ldots & \ldots & \ldots \\ 0\ \ 0 & 0 & \ldots 1 \\ 1\ \ 0 & 0 & \ldots -1 \end{bmatrix} \tag{3.11}$$

When the applied power supply is only U_{S1} in series between nodes U_1 and U_n, the equivalent inductance of the coil can be obtained from L_{eq} from formula (3.12), where θ_L is the impedance angle of the coil.

$$L_{eq} = \frac{U_{S1}}{\omega I_1}\sin(\theta_L) \tag{3.12}$$

Observing the distribution parameter model III of the coil, the energy storage unit of the coil has not only inductance but also parasitic capacitance. Similarly, the loss of the coil should include not only the copper loss of the coil but also the dielectric loss of the capacitor and the magnetic loss of the core. It is assumed that the coil does not have a core. According to the definition of quality factor, the ratio of stored energy to the consumed energy in a period, the quality factor Q_{L_1} of the coil is shown in formula (3.13), where S_L is the sum of the apparent power of each turn of the coil, and P_{LR} is the active power consumed by the coil, and the apparent power and active power include two parts: inductance and capacitance. $Q_{C(p,q)}$ is the quality factor of the parasitic capacitance between the p-segment and q-segment coils, Q_{Lm} is the quality factor of the m-segment coil, and Q_{Lm} and $Q_{C(p,q)}$ satisfy formula (3.14). The coil equivalent quality factor Q_{L_2} is measured by an impedance analyzer as shown in formula (3.15). The coil equivalent quality factor is equal to the absolute value of the tangent of the port input impedance angle θ_L, and $Q_{L\_1} = Q_{L\_2}$.

$$Q_{L\_1} = \frac{S_L}{P_{LR}} = \left| \frac{\sum_{m=1}^{n} I_m U_{Lm} + \sum_{p=1}^{n}\sum_{q=1}^{p-1} I_{C(p,q)}(U_p - U_q)}{\sum_{m=1}^{n} \frac{I_m U_{Lm}}{Q_{Lm}} + \sum_{p=1}^{n}\sum_{q=1}^{p-1} \frac{I_{C(p,q)}(U_p - U_q)}{Q_{C(p,q)}}} \right| \tag{3.13}$$

$$\begin{cases} Q_{Lm} = \frac{\omega L'_m}{R'_{Lm}} \\ Q_{C(p,q)} = \frac{1}{\omega C_{(p,q)} R_{C(p,q)}} \end{cases} \tag{3.14}$$

$$Q_{L\_2} = |\tan(\theta_L)| \tag{3.15}$$

When the operating frequency is low, the capacitive reactance $Z_{C(p,p)}$ of the parasitic capacitance is relatively large and the shunt effect on the coil is small. Therefore, it has little effect on the circuit and the current of the coil is satisfied formula (3.16). However, it is necessary to increase the operating frequency or reduce the coil turn spacing and increase the number of turns to improve the power transmission capability because of the relatively low coupling coefficient in IPT systems with small coil sizes. Therefore, the capacitive reactance $Z_{C(p,p)}$ of the parasitic capacitance is reduced, and the shunt effect on the coil cannot be ignored so that formula (3.16) is no longer established. The current difference between two adjacent inductor branches is shunted by the parasitic capacitance of the coil.

$$I_1 \approx I_2 \approx \cdots \approx I_n \tag{3.16}$$

3.1.3 Simulation and Experimental Verification

A double-layer 30-turn coil model is established in the simulation software, and the structure and dimensions are shown in Fig. 3.4. The outer diameter of the coil is 501.8 mm, the inner diameter is 383.4 mm, and the thickness is 22 mm. Each layer has 15 turns of coils, single turn coil diameter 3.2 mm, and coil turn spacing 0.8 mm. It is assumed that the winding sequence of the coil is from the outside of the upper layer to the inside and then from the inside to the outside of the lower layer. According to the winding order, each turn of coil is numbered 1–30, and the coil is divided into 30 segments. The coil model consists of two parts, copper wire and support structure, which is made of resin with a relative permittivity of 3.78.

According to the above model simulation, the self-inductance and mutual inductance distribution of each turn coil are shown in Fig. 3.5a, and the parasitic capacitance distribution is shown in Fig. 3.5b. The number of turns corresponding to the mutual inductance and parasitic capacitance is the number of each turn coil. For example, point A in Fig. 3.5b is the parasitic capacitance between turn 1 and turn 30. Figure 3.5 are symmetrical along a 45° angle. The inductance with the same abscissa and slave numbers is the self-inductance of the coil and the parasitic capacitance corresponding to the same number is equal to zero. The inductance with the same abscissa number and ordinate number is the self-inductance of the coil. The parasitic capacitance corresponding to the same number is equal to zero. Observing the parasitic capacitance distribution of the coil, it is found that the value of the adjacent two turns of the coil in the same layer is relatively large, above 100 pF. The total capacitance between the two layers of coils, C_{layer}, is equal to the sum of the capacitances between the two turns of the coils of different layers, so C_{layer} is equal to the sum of the capacitance values in the red box of Fig. 3.5b.

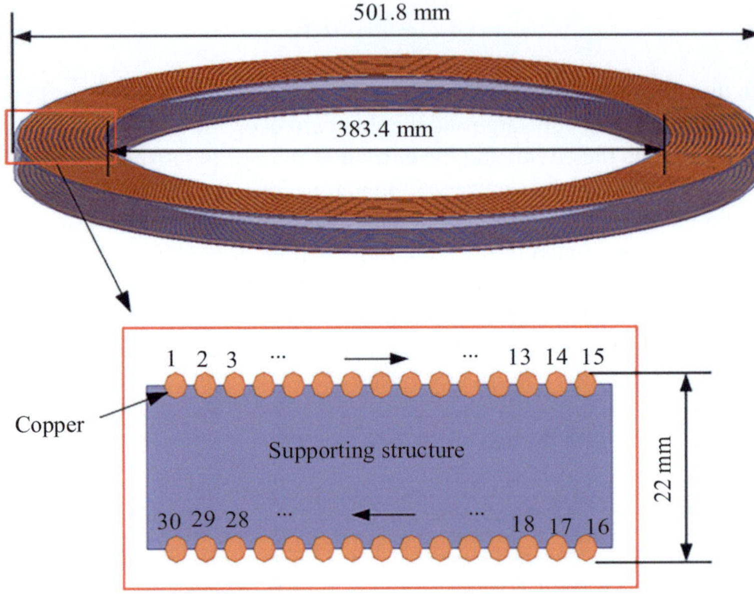

Fig. 3.4 Simulation model of a double-layer and 30-turn coil

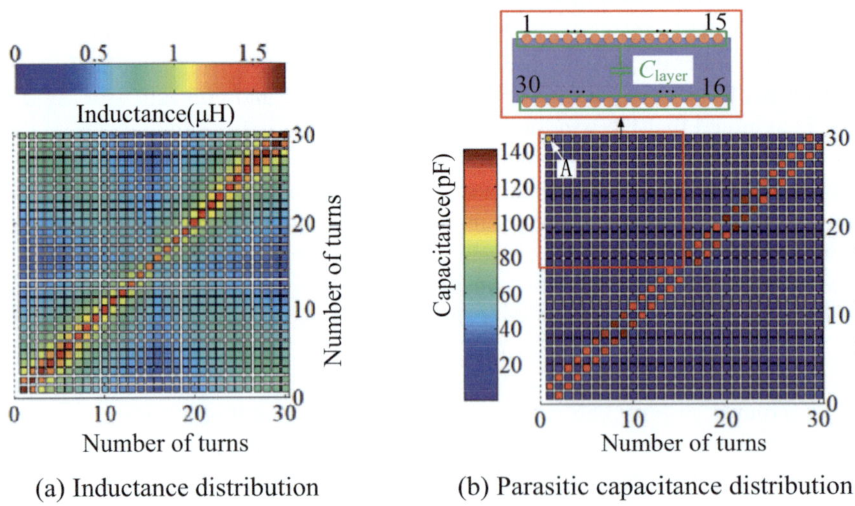

Fig. 3.5 Simulation results of inductance and parasitic capacitance distribution

Wind the coil according to the dimensions of Fig. 3.4. Since the parasitic resistance of the simulated wire per unit length is quite different from the measured value, the quality factor of the coil is calculated based on the measured value. The curve of the parasitic resistance of a 1-m wire varies with frequency as shown in Fig. 3.6. The quality factor of the parasitic capacitance in the simulation software is also not easy to obtain, and the quality factor of the coil is also calculated based on the measured value. The measurement results of the quality factor of C_{layer} at different frequencies are shown in Fig. 3.7. To simplify the calculation, assume that the quality factor of all parasitic capacitances is the same as C_{layer}.

From Fig. 3.5, the parameters of inductance and capacitance in the distributed parameter model can be calculated. Figures 3.6 and 3.7 can calculate the values of each resistor in the distribution parameter model. According to formula (3.12), the curve of the equivalent inductance of the coil L_{eq} varying with frequency is calculated, as shown in the red curve in Fig. 3.8. The curve of the measured equivalent inductor

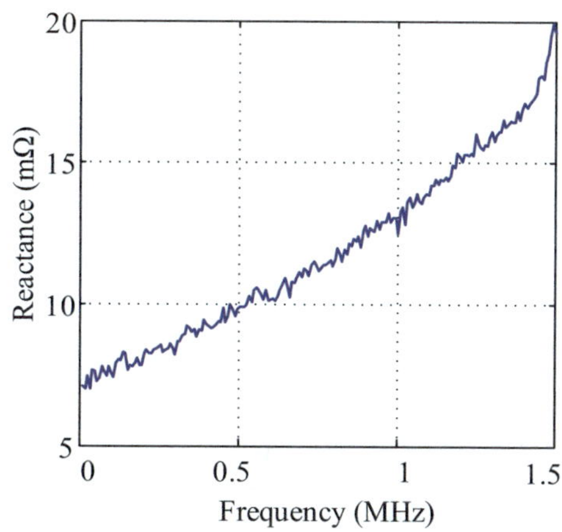

Fig. 3.6 The experimental results of parasitic resistance of 1-m wire

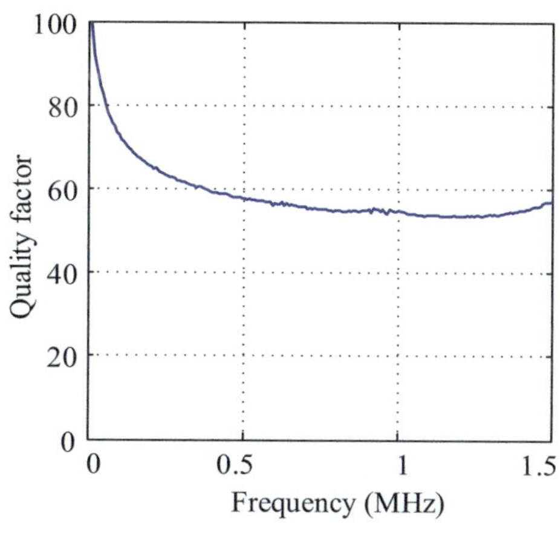

Fig. 3.7 The experimental results of the quality factor of parasitic capacitance

L_{eq} is shown in the blue curve. In the frequency range of 10 kHz–1.5 MHz, the coil equivalent inductance first increases with the increase of frequency and then decreases to a negative number. Large variations in the coil equivalent inductance are caused by parasitic capacitance. The calculation results show that the coil is self-resonant at a frequency of 0.687 MHz, and the experimental results show that the coil is self-resonant at a frequency of 0.771 MHz. The trend of theoretical calculation results and experimental results is basically the same, and the error is mainly caused by the inconsistency between the simulation and the actual parasitic capacitance values. There are three main causes of parasitic capacitance inconsistency. The first reason is that the wires used in the physical object are multi-stranded wires and are wrapped in multiple layers of insulating material. The second reason is that the coil support structure is doped with air bubbles during processing, resulting in an equivalent relative permittivity that does not correspond to the nominal value. The third reason is dimensional errors and parameter errors caused by the measuring wire head. The third reason is the error of coil size. When the maximum current of the coil is equal to 1 A, the current distribution of each turn coil at different frequencies is shown in Fig. 3.9. The higher the operating frequency, the greater the fluctuation in the distribution of coil current. The current difference between the two adjacent inductor branches is shunted by the parasitic capacitance of the coil. The low quality factor of the parasitic capacitance causes large losses, and the coil equivalent model deviates from the pure inductance.

The calculation results and experimental results of the coil quality factor are shown in Fig. 3.10. As the frequency increases, the quality factor first increases and then decreases to 0, and then increases.

The calculation results are basically the same as the experimental results, which proves the effectiveness of the proposed distribution parameter model and calculation method. Previous studies have ignored the dielectric loss of the parasitic capacitance itself, and the results of the coil quality factor calculation and experimental results without calculating the dielectric loss are shown in Fig. 3.11. The experiment and the calculation only match at low frequencies, and the error between the two increases

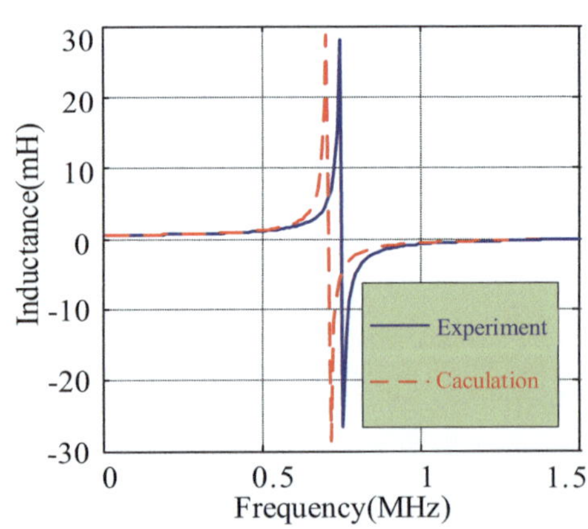

Fig. 3.8 Calculation and experimental result of the coil equivalent inductance

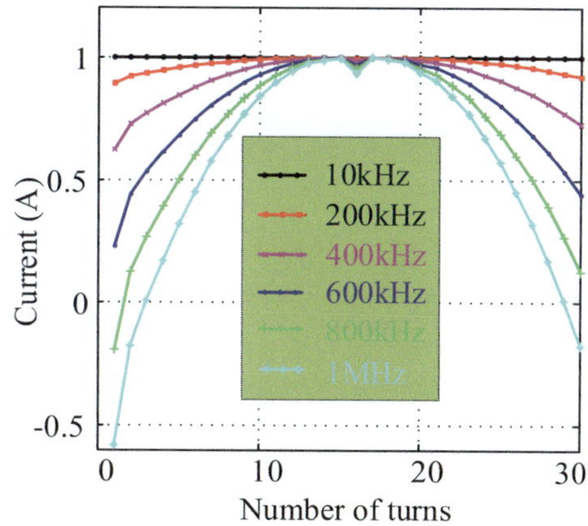

Fig. 3.9 Calculation results of the current distribution of each turn of the coil

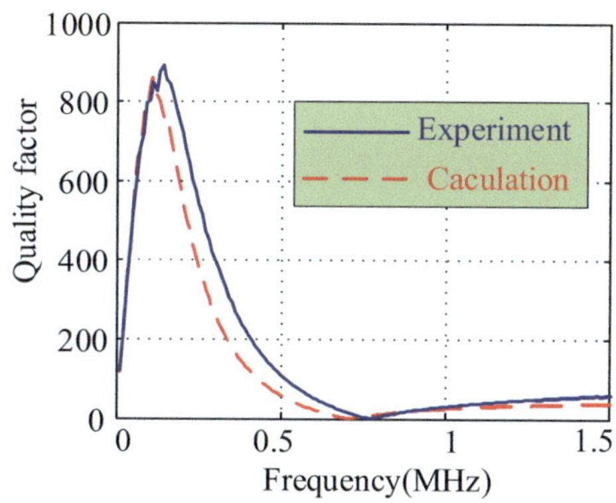

Fig. 3.10 Calculation and experimental result of the coil quality factor with the calculated dielectric loss

greatly as the frequency increases. The comparative results show that the parasitic capacitance dielectric loss of the distributed parameter model cannot be ignored.

3.2 Research on the Segmented Series Compensation Method of the Coil

In order to solve the problems of coil equivalent inductance change and quality factor reduction caused by parasitic capacitance in high-frequency wireless power transmission, a segmented series compensation method is proposed. Multiple compensation capacitors are connected in series with each section in series, and the parasitic capacitance is approximately "short-circuited". The voltage stress of the parasitic capacitance is effectively reduced, and the current flowing through the parasitic capacitance is also reduced. This method not only reduces the circulating current caused by the parasitic capacitance but also reduces the dielectric loss of the coil.

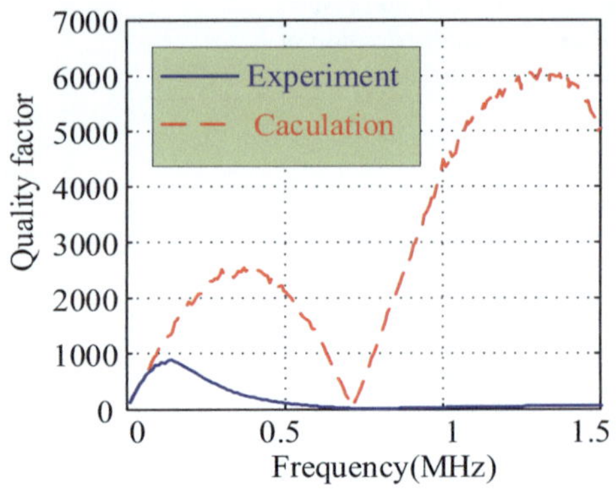

Fig. 3.11 Calculation and experimental result of the coil quality factor without the calculated dielectric loss

For different applications, two methods are proposed: single-layer segmented series compensation and single-turn segmented series compensation.

3.2.1 Single-Turn Segmented Series Compensation Method

(i) Model of a single-turn segmented series compensation coil

The single-turn segmented series compensation method is used for larger single-layer and multi-layer coils. Due to the large size of the coil, the inductance value of the single-turn coil is relatively large, and the parasitic capacitance value of the adjacent two-turn coil is also larger, so the shunt effect and high loss of the parasitic capacitance between turns cannot be ignored. Therefore, each turn of coil is compensated. Figure 3.12 is the schematic diagram of a planar spiral coil segmented compensation circuit with the compensation capacitor connected in series to the middle of each turn of the coil. Figure 3.13 is the circuit model of a single-turn segmented compensation coil. The parameters of each compensation capacitor are calculated by formula (3.17), where $L_{\mathrm{m\_turn}}$ is the equivalent inductance of the m-turn coil, which is equal to the sum of the self-inductance of the m-turn coil and the mutual inductance of this turn coil and other coils. The compensation capacitor resonates in series with one-half of the inductance value of the adjacent two-turn coil. jX_{A} is the reactance value required after compensation. Change X_{A} by adjusting the value of C_{L1}. When $X_{\mathrm{A}} = 0$, C_{L1} resonates in series with one-half of the sum of the inductances of the first and n-th turns coils. The voltage difference between any two points of U_{L1}, U_{L2}, …, U_{Ln} is almost zero. Therefore, the shunt effect of the parasitic capacitance can be ignored, so the coil can be equivalent to a model of a single inductor series capacitor. C_{A}, L_{A}, and R_{LA} are the total equivalent capacitor, equivalent inductor, and parasitic resistance respectively.

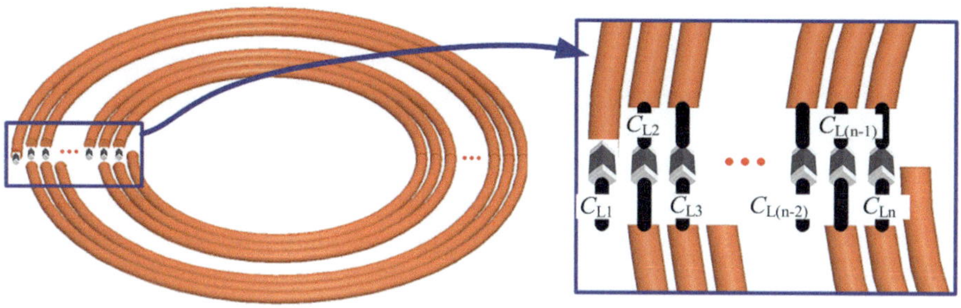

Fig. 3.12 Schematic diagram of segment compensation circuit of plane spiral coil

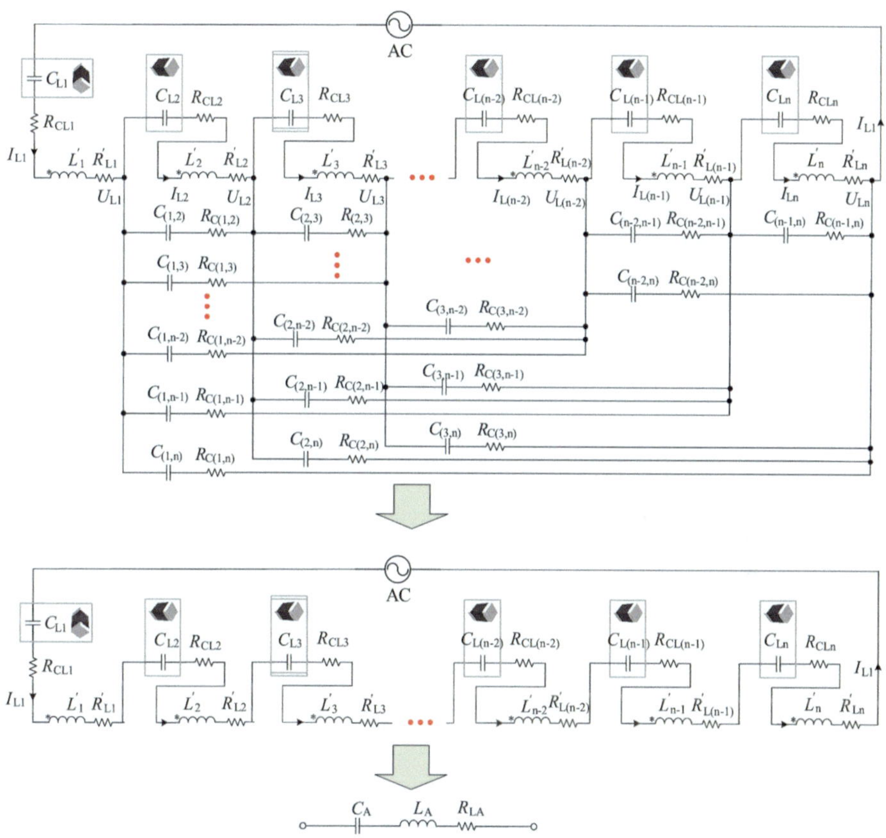

Fig. 3.13 Circuit model of the single-turn segmented compensation coil

The voltage and current distribution of the coil can be calculated by formula (3.4). Since the compensation capacitors are connected in series with the segments, the parameter matrix Z_L needs to be changed as shown in formula (3.18).

Planar spiral coils and solenoid coils are the two most commonly used large-size coil structures and are commonly used in long-range inductive IPT systems. The schematic diagram of the compensation capacitor connected to the planar spiral coil is shown in Fig. 3.12. The schematic diagram of the solenoid coil connection compensation capacitor is shown in Fig. 3.14. The compensation capacitor is connected in series with each turn of the coil, and the interturn voltage stress of the coil is compensated.

$$\begin{cases} C_{L1} = \frac{1}{-\omega X_A + \omega^2 L_1'} = \frac{1}{-\omega X_A + \omega^2 (L_{1\_turn} + L_{n\_turn})/2} \\ C_{Lm}|_{m=1,2,\ldots,n} = \frac{1}{\omega^2 L_m'} = \frac{2}{\omega^2 (L_{m-1\_turn} + L_{m\_turn})} \\ L_{Lm\_turn}\Big|_{m=1,2,\ldots,n} = L_m + \sum_{x=1}^{n(x \neq m)} M_{(m,y)} \end{cases} \tag{3.17}$$

$$Z_L = \begin{bmatrix} j\omega L_1' + R_{L1} + \frac{1}{j\omega C_{L1}} + R_{CL1} & j\omega M_{(1,2)}' & \cdots & j\omega M_{(1,n)}' \\ j\omega M_{(2,1)}' & j\omega L_2' + R_{L2}' + \frac{1}{j\omega C_{L2}} + R_{CL2} & \cdots & j\omega M_{(2,n)}' \\ \cdots & \cdots & \cdots & \cdots \\ j\omega M_{(n,1)}' & j\omega M_{(n,2)}' & \cdots & j\omega L_n' + R_{Ln}' + \frac{1}{j\omega C_{Ln}} + R_{CLn} \end{bmatrix} \tag{3.18}$$

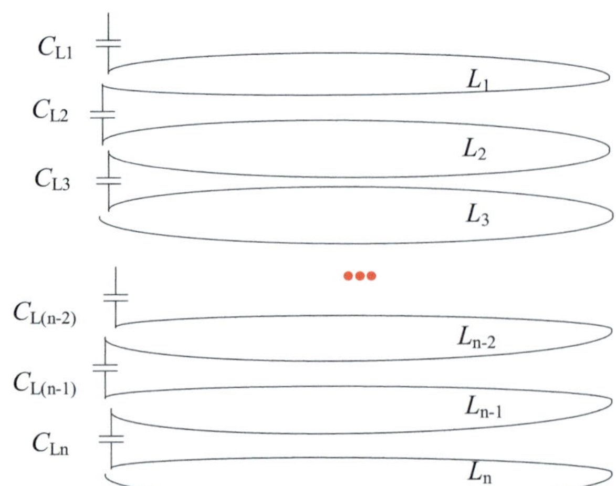

Fig. 3.14 Schematic diagrams of compensation capacitor connection for solenoid coil

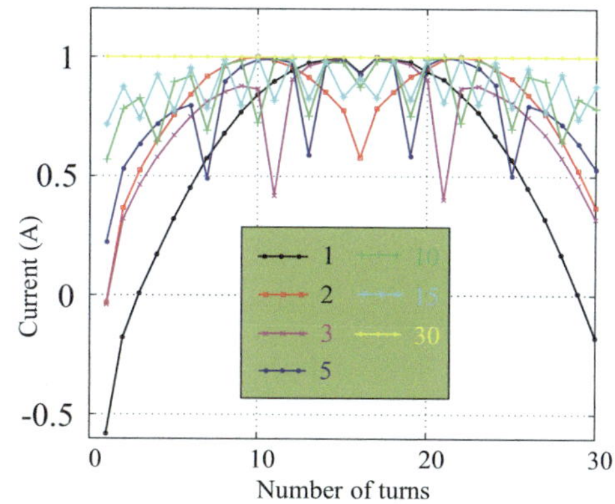

Fig. 3.15 Current distribution of the segment compensation coils

(ii) Simulation verification

The coil simulation model is shown in Fig. 3.4. The coils are divided into integer multiples of different turns of 1, 2, 3, 5, 6, 10, and 30. At a 1 MHz operating frequency, when the maximum current of the coil is equal to 1 A, the current distribution of each turn coil is shown in Fig. 3.15. The calculation results show that the more segments there are, the smaller the fluctuation of the current distribution of the coil.

When the coil is divided into 30 segments, the current per turn of the coil is almost the same. No current difference indicates that the current flowing through the parasitic capacitance is almost zero. The calculations show that the segmented compensation method can bypass the parasitic capacitance. Because the coil size is relatively large and the parasitic capacitance between turns is large, even if the two turns are divided into a segment, there is still a large current distribution fluctuation in the coil. Therefore, it is necessary to compensate for each turn, that is, single-turn segmented series compensation.

The quality factors of the parasitic capacitance and the parasitic resistance of 1-m wire are shown in Fig. 3.6 and Fig. 3.7 respectively. The quality factor of the compensation capacitor is 4875. When the coil is divided into different segments, the current distribution is shown in Fig. 3.15. The power and loss parameters of the coil are shown in Table 3.1, and the apparent power of the coil inductor P_{LS}, the apparent power of the parasitic capacitance P_{CS}, the copper loss of the coil P_{LR}, the dielectric loss of the parasitic capacitance P_{CR} and Q_{L_1} are calculated as shown in Eq. (3.19). P_{LS} reflects the magnetic energy stored by the coil, and P_{CS} reflects the electric field energy stored by the coil. With the increase of the number of coil segments, the ratio of P_{LR} to P_{CR} increases, the proportion of magnetic energy increases, and the coil changes from equivalent capacitive to equivalent inductive. The ratio of copper loss to dielectric loss also increases with the increase of the number of segments, from 1:161 for centralized compensation to 4712:1 for single-turn segmentation compensation. Centralized compensation is dominated by dielectric loss, and coil copper loss can be almost ignored. Single-turn segmentation compensation is dominated by copper loss, and dielectric loss can be almost ignored. The above calculations are obtained under the condition that the coil is unloaded. The nodal voltage difference of a single-turn segmented series compensation coil is mainly determined by the parasitic resistance of the coil inductor and the parasitic resistance of the compensation capacitor. The voltage difference of a single-turn segmented series compensation coil is mainly determined by the parasitic resistance of the coil inductor and the parasitic resistance of the compensation capacitor. When the coil is loaded, the voltage of the parasitic capacitance increases slightly.

$$\begin{cases} P_{\mathrm{LS}} = \sum\limits_{\mathrm{m}=1}^{\mathrm{n}} (I_{\mathrm{m}} U_{\mathrm{Lm}} - I_{\mathrm{m}}^2 (\frac{1}{\mathrm{j}\omega C_{\mathrm{m}}} + R_{\mathrm{Cm}})) \\ P_{\mathrm{CS}} = \sum\limits_{\mathrm{p}=1}^{\mathrm{n}} \sum\limits_{\mathrm{q}=1}^{\mathrm{p}-1} I_{C(\mathrm{p},\mathrm{q})} (U_{\mathrm{p}} - U_{\mathrm{q}}) \\ P_{\mathrm{LR}} = \sum\limits_{\mathrm{m}=1}^{\mathrm{n}} I_{\mathrm{m}}^2 R'_{\mathrm{Lm}} \\ P_{\mathrm{LC}} = \sum\limits_{\mathrm{p}=1}^{n} \sum\limits_{\mathrm{q}=1}^{\mathrm{p}-1} I_{\mathrm{C}(\mathrm{p},\mathrm{q})}^2 R_{\mathrm{C}(\mathrm{p},\mathrm{q})} \\ Q_{\mathrm{L\_1}} = \frac{S_{\mathrm{L}}}{P_{\mathrm{L}}} = \left| \frac{P_{\mathrm{LS}} + P_{\mathrm{CS}}}{P_{\mathrm{LR}} + P_{\mathrm{LC}}} \right| \end{cases} \tag{3.19}$$

Comparing the unsegmented centralized compensation and single-turn segmented compensation, the current distribution of the parasitic capacitance at 1MHz operating frequency with a maximum coil current of 1 A is shown in Fig. 3.16. The parasitic capacitance current of the centrally compensated coil can reach up to 0.177 A, and the high current inevitably causes high dielectric losses. The parasitic capacitance current of the centralized compensation coil is more than 1000 times that of single-turn segmented compensation, and the maximum parasitic capacitance current of the single-turn segmented compensation coil is only 1.05 mA. The loss is proportional to the square of the current, so the dielectric loss of the parasitic capacitance is almost negligible, which further demonstrates the effectiveness of single-turn segmentation compensation.

Table 3.1 Power and loss parameters of coils with different number of segments

Number of segments	P_{LS} (W)	P_{CS} (W)	P_{LR} (W)	P_{CR} (W)	$Q_{\mathrm{L\_1}}$ (W)
1	682.2	1326	0.1436	23.19	27.62
2	1238	935.0	0.1944	16.35	18.36
3	1038	858.5	0.1691	15.01	11.87
5	1237	617.4	0.1929	10.79	56.43
10	1444	285.3	0.2184	4.988	222.6
15	1505	153.9	0.2254	2.690	463.5
30	1977	3.56e−4	0.2938	6.235e−6	6730

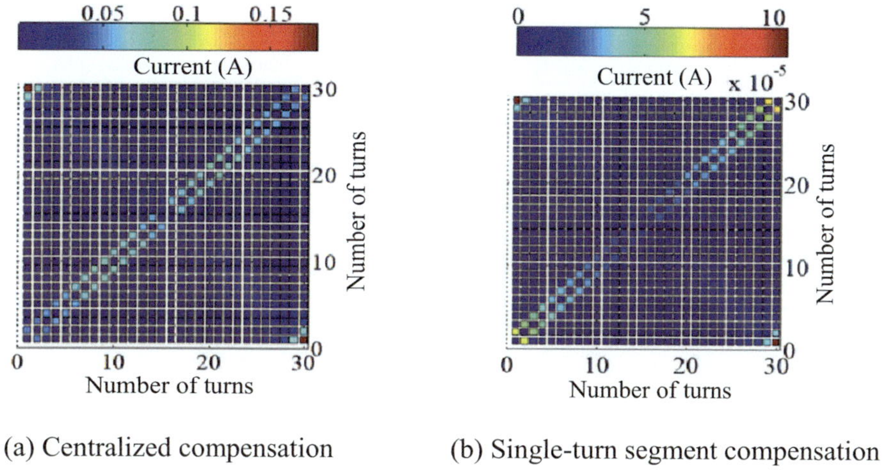

Fig. 3.16 Distribution diagrams of parasitic capacitance currents for lumped compensation and single-turn segmented compensation

(iii) Experimental verification

Wind two coils according to the structural dimensions of Fig. 3.4, one on the primary side and one on the secondary side. The system operates at 1 MHz. According to formula (3.17), the matching compensation capacitor is calculated and selected to resonate in series with the primary and secondary side coils. The parameters of the primary side compensation capacitor are shown in Table 3.2, and the parameters of the secondary side compensation capacitor are shown in Table 3.3.

The sum of the parasitic resistances of the primary compensation capacitor is R_{CP_2} 0.81 Ω, and the sum of the parasitic resistance of the secondary compensation capacitor is R_{CS_1} 0.82 Ω.

Table 3.2 The parameters of the primary compensation capacitor

Variable	Value (nF)	Variable	Value (nF)	Variable	Value (nF)
C_{L1p}	1.284	C_{L11p}	1.198	C_{L21p}	1.211
C_{L2p}	1.205	C_{L12p}	1.248	C_{L22p}	1.171
C_{L3p}	1.163	C_{L13p}	1.299	C_{L23p}	1.151
C_{L4p}	1.139	C_{L14p}	1.363	C_{L24p}	1.134
C_{L5p}	1.122	C_{L15p}	1.448	C_{L25p}	1.123
C_{L6p}	1.118	C_{L16p}	1.504	C_{L26p}	1.128
C_{L7p}	1.120	C_{L17p}	1.446	C_{L27p}	1.123
C_{L8p}	1.134	C_{L18p}	1.360	C_{L28p}	1.133
C_{L9p}	1.143	C_{L19p}	1.292	C_{L29p}	1.157
C_{L10p}	1.175	C_{L20p}	1.239	C_{L30p}	1.2

Table 3.3 The parameters of the secondary compensation capacitor

Variable	Value (nF)	Variable	Value (nF)	Variable	Value (nF)
C_{L1s}	1.280	C_{L11s}	1.201	C_{L21s}	1.207
C_{L2s}	1.207	C_{L12s}	1.248	C_{L22s}	1.176
C_{L3s}	1.158	C_{L13s}	1.289	C_{L23s}	1.133
C_{L4s}	1.139	C_{L14s}	1.360	C_{L24s}	1.131
C_{L5s}	1.127	C_{L15s}	1.453	C_{L25s}	1.123
C_{L6s}	1.122	C_{L16s}	1.497	C_{L26s}	1.123
C_{L7s}	1.12	C_{L17s}	1.42	C_{L27s}	1.126
C_{L7s}	1.137	C_{L18s}	1.363	C_{L28s}	1.134
C_{L9s}	1.154	C_{L19s}	1.292	C_{L29s}	1.163
C_{L10s}	1.171	C_{L20s}	1.242	C_{L30s}	1.198

Taking the primary coil as an example, the curve of reactance of the compensated coil varying with frequency is shown in Fig. 3.17. The reactance value of simulation and experiment at 1 MHz is almost zero. The simulation curve is not completely consistent with the experiment. The slight error is mainly due to the difference in parasitic capacitance. The curve of parasitic resistance of the compensated coil varying with frequency is shown in Fig. 3.18. The parasitic resistance of the compensated coil at 1 MHz is the smallest. The above experimental results show that the single-turn segmented compensation coil achieves series resonance at a frequency of 1 MHz.

The measured value of the parasitic resistance of the single-turn segmented series compensation coil R_{par_E} at 1 MHz is 1.95 Ω, and the calculated value R_{par_C} is 1.4 Ω, and the difference is mainly caused by the proximity effect. In the uncompensated coil, the loss mainly contains the copper loss of the coil and the dielectric loss of the parasitic capacitance. Copper loss can be divided into skin-effect copper loss and proximity-effect copper loss. Because the copper loss caused by proximity effect and the dielectric loss of parasitic capacitance are both related to the distance between each turn of the coil, they cannot be divided before the coil is compensated.

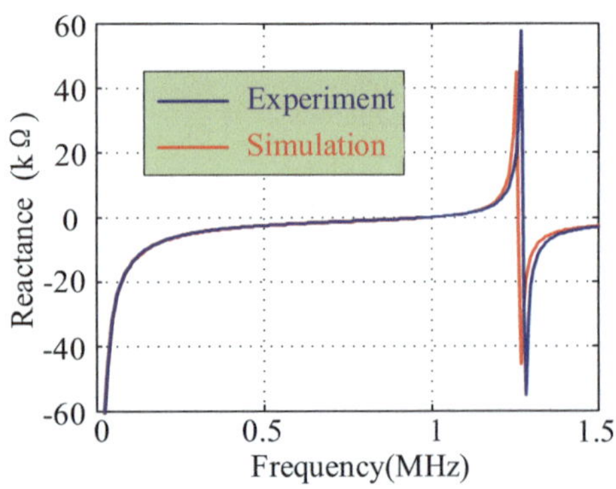

Fig. 3.17 The curve of the reactance of the single-turn segmented compensation coil varying with frequency

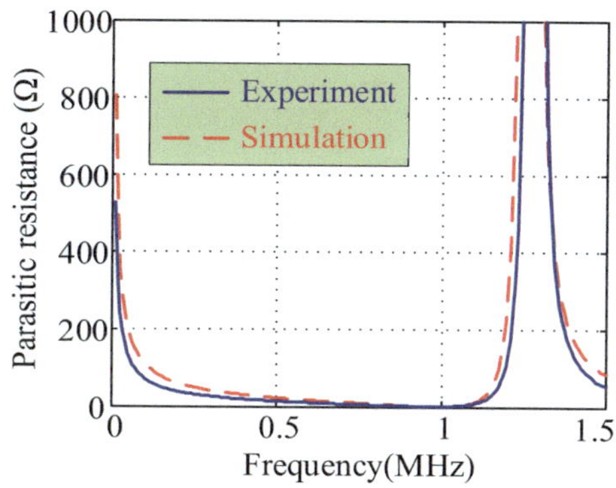

Fig. 3.18 The curve of the parasitic resistance of the single-turn segmented compensation coil varying with frequency

At the operating frequency, the dielectric loss of the parasitic capacitance of the compensated coil is almost zero. Therefore, the loss of the compensated coil only includes the copper loss and the compensation capacitor loss. The parasitic resistance of the compensation capacitor in simulation is based on the measured value of the experiment, so they are the same. In simulation, the parasitic resistance of the inductor is calculated based on the measured parasitic resistance value of the 1-m coil. Therefore, the parasitic resistance of the inductor in the simulation only includes the copper loss caused by skin effect. In experiment, the parasitic resistance of the inductor reflects both copper loss caused by skin effect and proximity effect. Therefore, the difference between R_{par_E} and R_{par_C} is the proximity effect copper loss. The loss distribution of the compensation coil is shown in Fig. 3.19.

A four-coil IPT system with a high pitch-to-diameter ratio was built. The two large coils wound serve as repeater coils, and their main function is to establish a strong magnetic field and realize long-distance wireless power transmission. Two other small coils are used for transmitting and receiving, and the experimental prototype is shown in Fig. 3.20. The circuit schematic of the four-coil IPT system is shown in Fig. 3.21. Q_1 and Q_2 are the two MOSFETs of the half-bridge inverter, C_1 and C_2 are the regulated capacitors of the half-bridge inverter, L_{P1} is the primary transmitting coil, L_{P2} is the primary relay coil, L_{S1} is the secondary side relay coil, L_{S2} is the secondary receiving coil, and the C_{P_1}, C_{P_2}, C_{S_1}, and C_{S_2} are the system

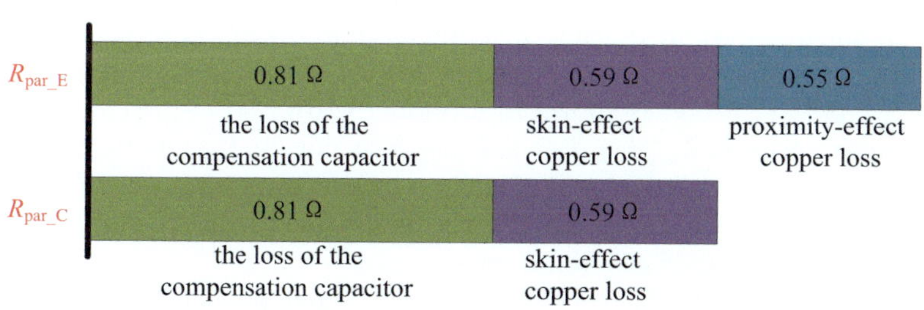

Fig. 3.19 Experimental and calculated loss distribution of compensation coil

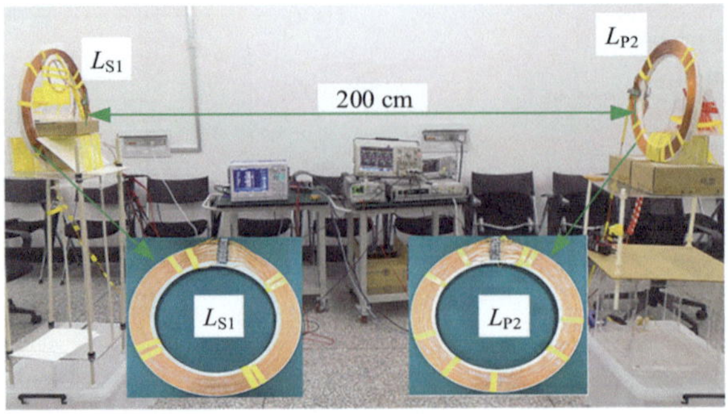

Fig. 3.20 Experimental prototype of single-turn segmented series compensation

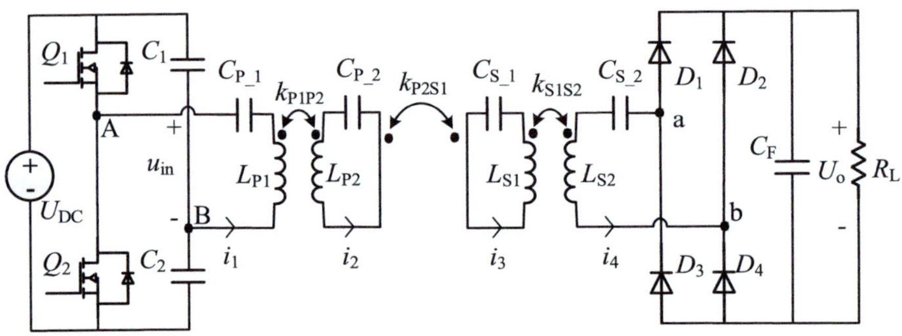

Fig. 3.21 Circuit schematic of four-coil IPT system

compensation capacitors, which resonate in series with L_{P1}, L_{P2}, L_{S1}, and L_{S2}, respectively. C_{P_2} and C_{S_1} are equivalent capacitance values for 30 series capacitors on the primary and secondary sides, respectively. D_1 ~ D_4 are the four diodes of the bridge rectifier, C_F is the filter capacitor, and R_L is the load resistance.

When the transmission distance is 2 m, the circuit parameters are shown in Table 3.4. Because the value of k_{P2S1} is so small that it is almost impossible to measure, the value of k_{P2S1} is obtained by simulation. Compared with L_{P2} and L_{S1}, the inductance of coils L_{P1} and L_{S2} is relatively small, and k_{P2S1} is much smaller than k_{P1P2} and k_{S1S2}, so the transmission efficiency $\eta_{P2\text{-}S1}$ from L_{P2} to L_{S1} basically determines the efficiency of the system. When the coil is fully compensated, the efficiency from L_{P2} to L_{S1} is shown in formula (3.20). R_{P1}, R_{P2}, R_{S1}, and R_{S2} are parasitic resistances for L_{P1}, L_{P2}, L_{S1}, and L_{S2}, respectively. R_{P2} and R_{S1} are equal to 1.14 Ω and 1.16 Ω, respectively. The quality factors Q_{LP2} and Q_{LS1} of L_{P2} and L_{S1} are equal to 3464 and 3407, respectively. The quality factor Q_{CP_2} and Q_{CS_1} of C_{P_2} and C_{S_1} are equal to 4875 and 4820, respectively.

The curves of system efficiency (DC/DC) $\eta_{dc\text{-}dc}$ and $\eta_{P2\text{-}S1}$ with load resistance are shown in Fig. 3.22. $\eta_{DC\text{-}DC}$ is the experimental value, and $\eta_{P2\text{-}S1}$ is calculated according to Eq. (3.20). When R_L is equal to 75 Ω, the system efficiency $\eta_{DC\text{-}DC}$ is

Table 3.4 Circuit parameters when the transmission distance is 2 m

Parameter	Value	Unit	Parameter	Value	Unit
f	1	MHz	L_{S1}	629.0	μH
U_{in}	88	V	L_{S2}	1.54	μH
k_{P1P2}	0.038	–	C_{P_1}	18.53	nF
k_{P2S1}	0.0008	–	C_{P_2}	40.30	pF
k_{S1S2}	0.087	–	C_{S_1}	40.27	pF
L_{P1}	1.39	μH	C_{S_2}	15.11	nF
L_{P2}	628.5	μH	R_L	75	Ω

the highest. The power transfer at maximum efficiency is shown in Fig. 3.23, with input and output power of 396.9 W and 115.2 W, respectively, and a system efficiency of 29%. Figure 3.24 is the waveform of the output voltage of the inverter, the output current of the inverter, the current of the primary relay coil, and the output voltage of the system. The output voltage of the inverter slightly leads the output current, and the input impedance is weak inductive. Figure 3.25 is the waveforms of current of the primary and secondary relay coils. The current of the primary relay coil is 16 A, and the current of the secondary relay coil is 6.5 A. The current of the primary coil is lag 90° compared to the secondary coil.

$$\begin{cases} \eta_{\text{P2-S1}} = \frac{\alpha}{1+\alpha} \times \frac{k_{P2S1}^2 Q_{P2} Q_{S1}}{k_{P2S1}^2 Q_{P2} Q_{S1} + 1 + \alpha} \\ \alpha = \frac{k_{S1S2}^2 \omega^2 L_{S1} L_{S2}}{R_{S1}(R_{re} + R_{S2})} \\ Q_{P2} = \frac{Q_{LP2} Q_{CP\_2}}{Q_{LP2} + Q_{CP\_2}} \\ Q_{S1} = \frac{Q_{LS1} Q_{CS\_1}}{Q_{LS1} + Q_{CS\_1}} \end{cases} \tag{3.20}$$

Figure 3.26 is the input and output power measured by the power analyzer. The loss of L_{P2}, C_{P_2}, L_{S1}, and C_{S_1} accounted for 96.6% of the total loss of the system, and the

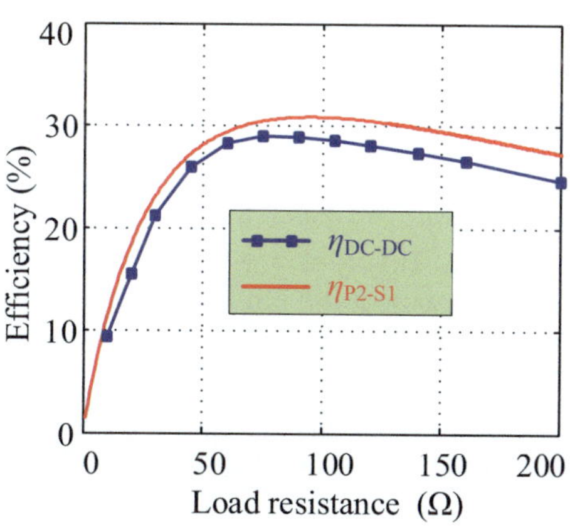

Fig. 3.22 Curves of η_{DC-DC} and η_{P2-S1} varying with load resistance

Fig. 3.23 Power transfer diagram at maximum efficiency

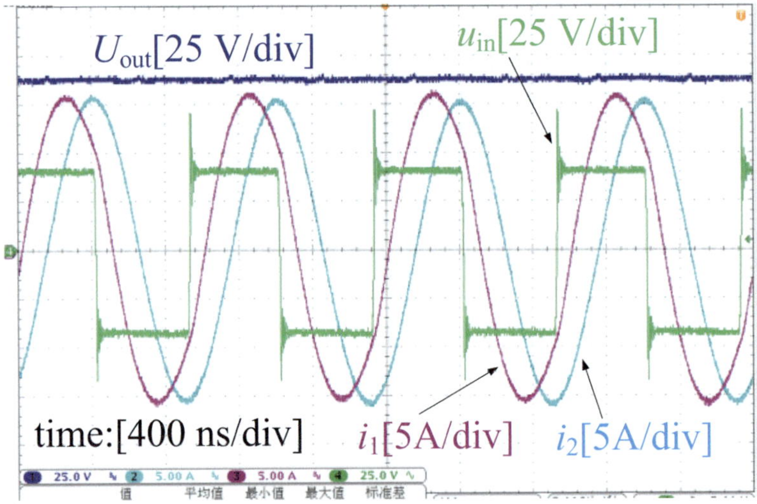

Fig. 3.24 Waveforms of inverter output voltage and current, primary relay coil current and system output voltage

transmission efficiency of the two relay coils basically determined the transmission efficiency of the system, which is consistent with the previous analysis. Experimental results show that the proposed single-turn segmented series compensation method can effectively reduce the dielectric loss of tightly wound large-size coils, which is helpful to improve the transmission efficiency and power density of the system. The coil with single-stage compensation is capacitive at 1 MHz, as shown in Fig. 3.8. The measured value of the inductance is −783 μH, and the parasitic resistance is 125 Ω.

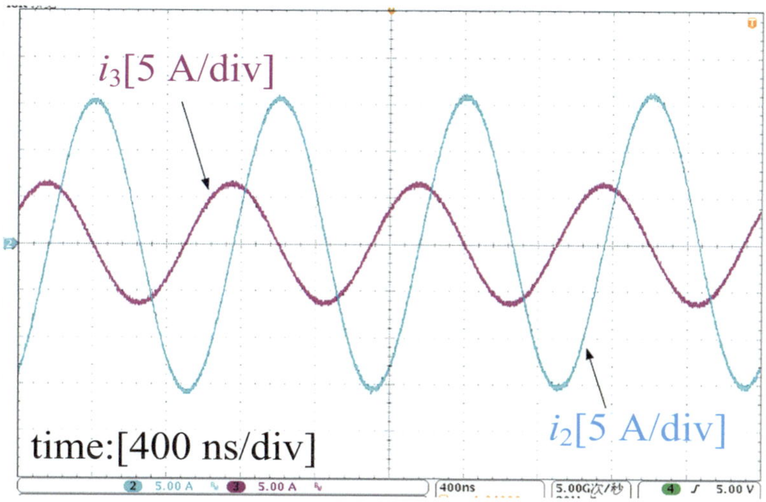

Fig. 3.25 Current waveforms of primary and secondary relay coils

Therefore, it cannot be used to transmit energy to compare with the improved multi-stage compensation. In order to prove the advantages of the single-turn segmented series compensation method, the prototype is compared with the previous research work with high-pitch-diameter ratio IPT system, and the comparison parameters are shown in Table 3.5. In previous studies, the coil could not be tightly wound because the parasitic capacitance problem was not solved, so the coil size was larger and the system efficiency was lower. It can be seen from formula (3.20) that under the condition of weak coupling, the transmission efficiency is approximately proportional to the square of the coupling coefficient. Although the coupling coefficient of the system designed in this paper is much lower than that of previous research work, the transmission efficiency is still as high as 29%, which is the reason for the high coil quality factor. The comparative results show the important role of the proposed method in improving the efficiency of the system.

3.2.2 Single-Layer Segmented Series Compensation Method

(i) Model of a single-layer segmented series compensation coil

Single-layer segmented series compensation is suitable for small size multi-layer coils. Due to the relatively small size of the coil, the parasitic capacitance between two adjacent turns of the same layer is relatively small, and the impact on the circuit is negligible. However, the total parasitic capacitance between the layers is large, and the voltage stress is relatively large, so its influence on the coil cannot be ignored. Therefore, each layer of coil is individually compensated. Figure 3.27 is a schematic

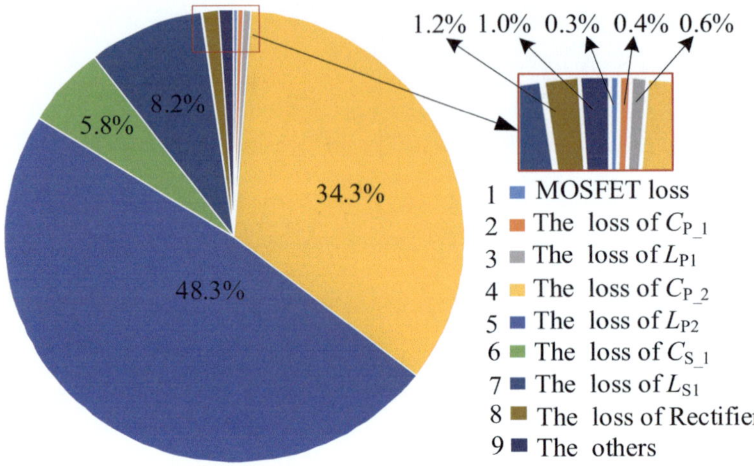

Fig. 3.26 Loss distribution diagram of the system

Table 3.5 Circuit parameters of several high distance-to-diameter ratio wireless power transmission systems

Reference	Frequency (MHz)	Size (cm)	Coupling coefficient	Quality factor	Output power (W)	Efficiency (%)
49	0.15	50 × 50 × 19.2	88e−4	1600	103	75
53	10	ϕ60 × 20	18e−4	950	60	15
54	7.64	ϕ59 × 0.254	68e4	×××	12	50
This paper	1	ϕ50.2 × 2.2	8e−4	3464	115	29

diagram of the segmented compensation circuit of the b-layer and n-turn coils. The number of turns per layer is $d = n/b$. In addition to the capacitance being connected in series between the coil layers, the single-layer segmented series compensation also requires the winding method of the coil, and the winding order and direction of each layer of coil need to be consistent, as shown in Fig. 3.20. Each layer of coil is wound counterclockwise from the outside to the inside, and the traditional coil winding method usually adopts the winding method of the adjacent layers in reverse order and in the same clockwise direction to avoid wire crossing. The parameter design method of the compensation capacitor is shown in formula (3.21). In addition to C_{L1}, the other compensation capacitors resonate with one-half of the sum of the inductances of the two turns of coils in the adjacent layer. L_{m_layer} is the inductance of the m-layer coil, and jX_A is the reactance value required after segmented compensation. By adjusting the value of C_{L1}, the equivalent reactance value of the coil after compensation is equal to jX_A. When $X_A = 0$, C_{L1} resonates in series with one-half of the sum of the inductances of the first and b-th coils.

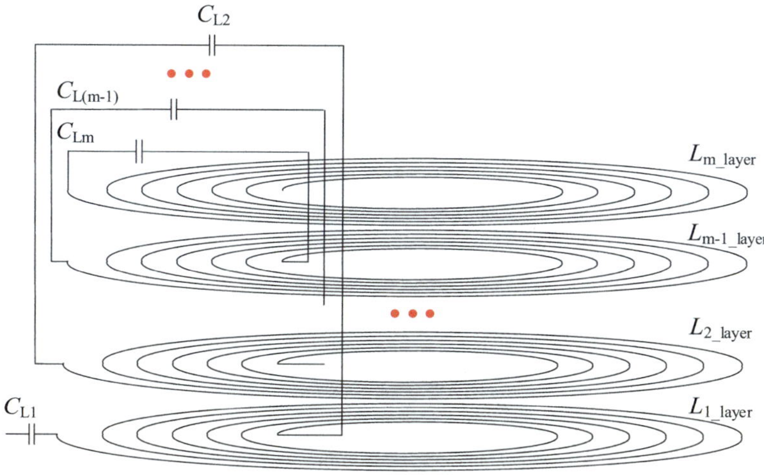

Fig. 3.27 Single-layer segmented series compensation circuit of b-layer and n-turn coil

$$\begin{cases} C_{L1} = \dfrac{1}{-\omega X_A + \omega^2 (L_{1\_layer} + L_{n\_layer})} \\ C_{L(dm+1)}\big|_{m=2,\ldots,b-1} = \dfrac{2}{\omega^2 (L_{m-1\_layer} + L_{m\_layer})} \\ L_{m\_layer}\big|_{m=1,2,\ldots,b} = \sum\limits_{x=1}^{d} L_{dm+x} + \sum\limits_{x=1}^{d} \sum\limits_{y=1}^{n(y\neq x)} M_{(x,y)} \end{cases} \tag{3.21}$$

Figure 3.28 shows the circuit model of a single-layer segmented compensation coil. The coil is compensated by a single layer segment, the voltage difference between the adjacent two turns of the coil between the layers is close to zero, and the current flowing through the parasitic capacitance is also very small, and the parasitic capacitance is bypassed. The nodal voltage and branch current of the coil can be calculated by formula (3.4). Because the compensation capacitors are connected in series, change the parameter matrix Z_L as shown in formula (3.22). Z_{Lm1} is the reactance value of the m_1 turn coil, Z_{CLm2} is the reactance value of the m_2 compensation capacitor, and the calculation is shown in formula (3.23).

$$Z_L = \begin{bmatrix} Z_{L1}+Z_{CL1} & \cdots & j\omega M'_{(1,d+1)} & \cdots & j\omega M'_{(1,n-d+1)} & \cdots & j\omega M'_{(1,n)} \\ \cdots & \cdots & \cdots & \cdots & \cdots & \cdots & \cdots \\ j\omega M'_{(d+1,1)} & \cdots & Z_{L(d+1)}+Z_{CL2} & \cdots & j\omega M'_{(d+1,n-d+1)} & \cdots & j\omega M'_{(d+1,n)} \\ \cdots & \cdots & \cdots & \cdots & \cdots & \cdots & \cdots \\ j\omega M'_{(n-d+1,1)} & \cdots & j\omega M'_{(n-d+1,d+1)} & \cdots & Z_{L(n-d+1)}+Z_{CLb} & \cdots & j\omega M'_{(n-d+1,n)} \\ \cdots & \cdots & \cdots & \cdots & \cdots & \cdots & \cdots \\ j\omega M'_{(n,1)} & \cdots & j\omega M'_{(n,d+1)} & \cdots & j\omega M'_{(n,n-d+1)} & \cdots & Z_{Ln} \end{bmatrix} \tag{3.22}$$

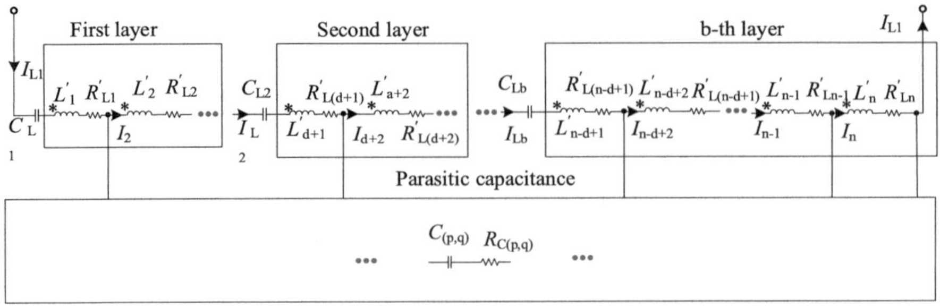

Fig. 3.28 Circuit model of the single-layer segmented compensation coil

$$\begin{cases} Z_{\mathrm{Lm1}} = \mathrm{j}\omega L'_{\mathrm{m1}} + R'_{\mathrm{Lm1}} \\ Z_{\mathrm{CLm2}} = \frac{1}{\mathrm{j}\omega C_{\mathrm{Lm2}}} + R_{\mathrm{CLm2}} \end{cases} \tag{3.23}$$

Taking a 2-layer 10-turn coil as an example, the schematic diagram of the comparison between the improved and the traditional method is shown in Fig. 3.29. The outlet end and the inlet end of the coil wound by the improved method are on the same side. The voltage difference between layers increases sequentially from inside to outside, and the amplitude of the red sine wave in Fig. 3.29 indicates the magnitude of the voltage. The outlet end and the inlet end of the winding coil are on the inside and outside respectively, and the voltage between the layers is almost equal. The coil wound by the improved method is compensated in segments by a single layer, and the parasitic capacitance between layers is approximated to be short-circuited, and its voltage is almost zero. The comparison of the three methods shows that the improved winding and compensation methods can reduce the dielectric loss of the coil.

(ii) Simulation verification

The LCT model of a pair of two-layer 34-turn coils is shown in Fig. 3.30. The outer diameter of the coil is 175 mm, the inner diameter is 83.5 mm, the wire diameter is 2.5 mm, and the layer spacing is 0.2 mm. The transmission distance is 71 mm.

The self-inductance, mutual inductance, and parasitic capacitance of the coil are obtained by simulation. Then we can calculate the voltage between two adjacent turns of the adjacent layer at 100 kHz. When the current of the coil is 1 A, the curve of the voltage difference between two adjacent turns of coils in different layers from inner to outer is shown in Fig. 3.31. Compared with the traditional method, the improved winding and compensation method greatly reduces the voltage of the parasitic capacitors between layers. Because the parasitic capacitance value of the adjacent coil is relatively large and the quality factor is very low, high voltage means high loss. Therefore, the coil loss caused by parasitic capacitance is greatly reduced, which will be very helpful to improve system efficiency.

Figure 3.32 shows the curves of coils' self-inductance wound by traditional and improved methods varying with frequency from 100 Hz to 2 MHz. The self-resonant

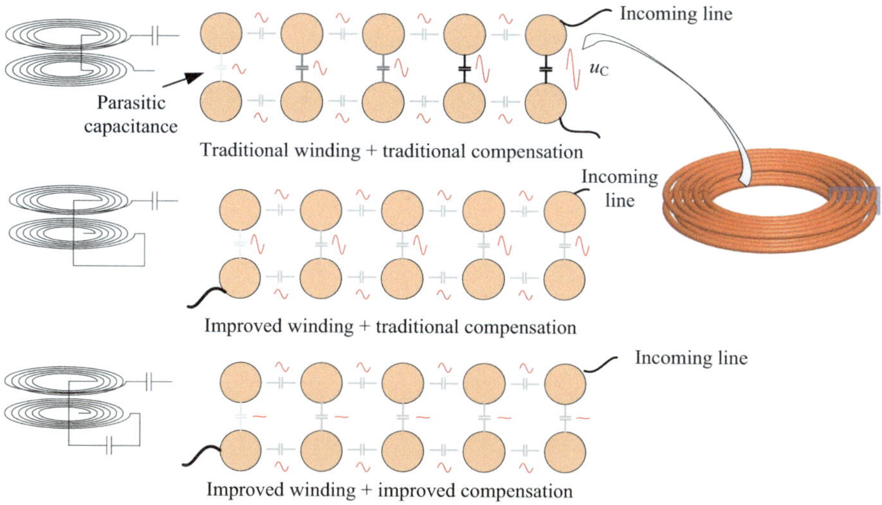

Fig. 3.29 Comparison of the improved and traditional method of a 2-layer and 10-turn coil

Fig. 3.30 LCT model consisting of two two-layer and 34-turn coils

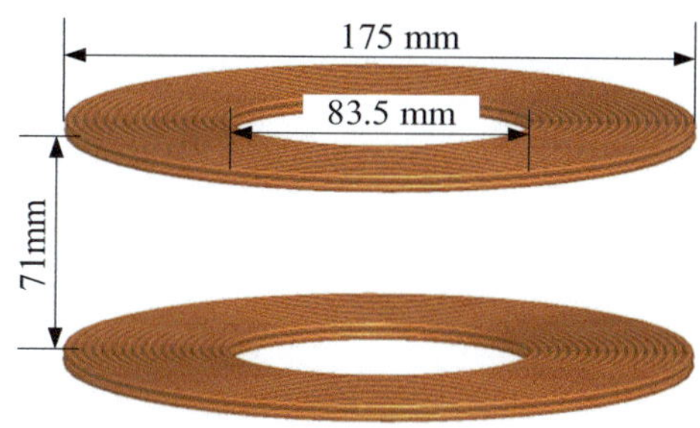

frequencies of the two coils are 1.14 MHz and 1.38 MHz, respectively, which is due to the influence of parasitic capacitance between layers. The self-resonant frequency of the coil in improved winding method is higher, which indicates the less influence of its parasitic capacitance. Along the winding sequence of the coil, the current distribution diagram of the double-layer 34-turn coil is shown in Fig. 3.33. The greater the voltage between the layers, the greater the current difference between the adjacent two turns of the coil. The uneven distribution of coil current is caused by the shunt of parasitic capacitance. Compared to conventional methods, the coil current using the improved winding and compensation method is almost unchanged.

(iii) Experimental verification

Wind two LCTs according to the parameters shown in Fig. 3.30. One is wound using the traditional method, and the other is wound using the improved method. The curve of the equivalent inductance of the coil varying with frequency is shown in Fig. 3.34. The self-resonant frequency of the winding coil by the traditional method is 1.14

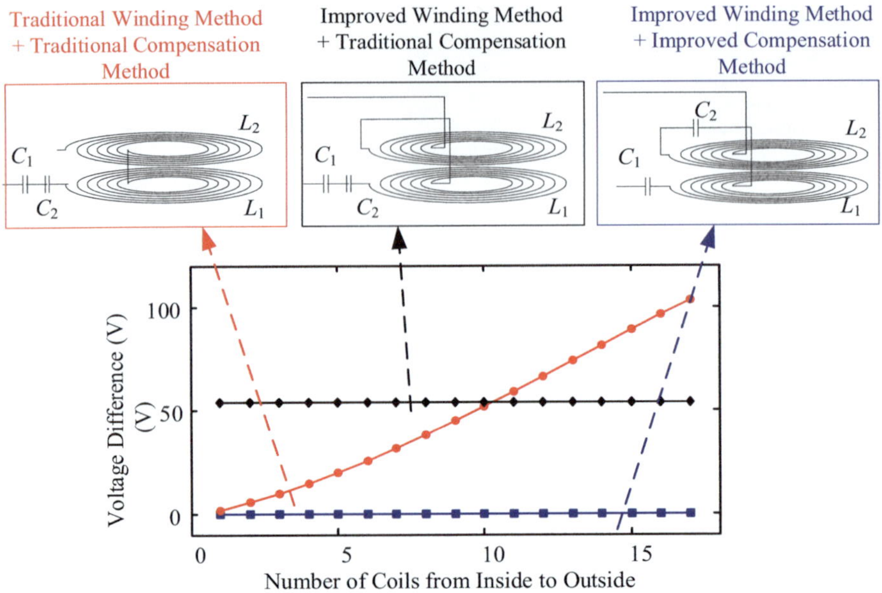

Fig. 3.31 Interlayer voltage under different winding and compensation methods

Fig. 3.32 Calculation results of the equivalent inductance of the coil

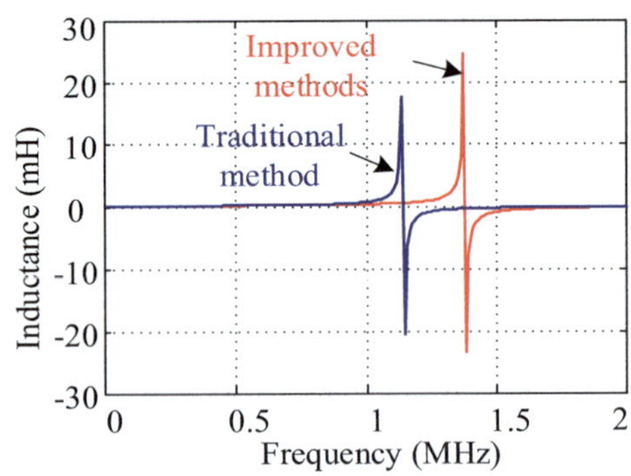

MHz, and the self-resonant frequency of the winding coil of the improved method is 1.34 MHz, and the experimental results are basically consistent with the simulation.

The traditional winding and compensation method and the improved winding and compensation method were used to build the system prototype. Except for the LCT and its series compensation capacitors, the other parameters are the same. The prototype is shown in Fig. 3.35a, the coil is shown in Fig. 3.35b, and the components in the red box are the compensation capacitors connected in series between the first and second layers. The system circuit schematic is shown in Fig. 3.36, the compensation topology is LCC/S, and its circuit parameters are shown in Table 3.6. The system operates at 100 kHz. The coupling coefficient of LCT winding by traditional

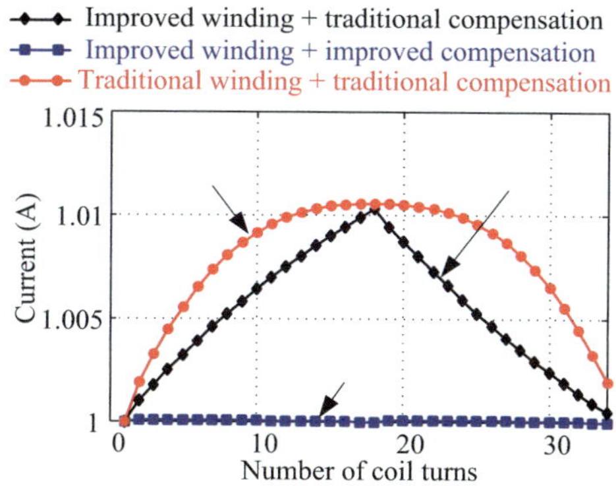

Fig. 3.33 Current distribution diagram of the two-layer and 34-turn coil

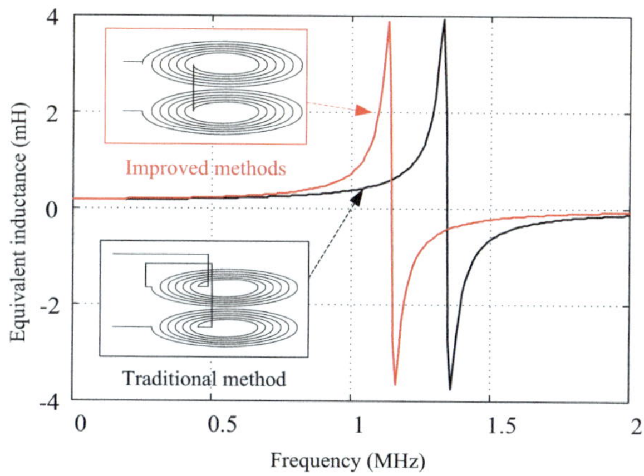

Fig. 3.34 Experimental results of the equivalent inductance of the coil wound by the improved and methods

method is 0.154, and the coupling coefficient of LCT winding by improved method is 0.152. L_{P1} is the primary series compensation inductor, C_{P1} is the primary parallel compensation capacitor, C_{P2} is the primary series compensation capacitor, C_{S1} is the secondary series compensation capacitor, L_P and L_S are the primary, and the secondary coil self-inductance, respectively. The symbol with "tra" subscript indicates the parameter of the traditional method, and the symbol with "im" subscript indicates the parameter of the traditional method. C_{p2_im} is composed of C_{p2_im1} and C_{p2_im2}, and C_{s1_im} is composed of C_{s1_im1} and C_{s1_im2}. C_{p2_im2} and C_{s1_im2} are connected in series between the two layers of the primary and secondary coils, respectively. The size, the number of turns and the relative position of the traditional coil, and the improved coil are under the same condition, so their parameters are also almost equivalent.

Table 3.7 lists the parasitic resistance of the LCT and its series compensation capacitors where "$L_{P\_tra} + C_{P\_tra}$" and "$L_{S\_tra} + C_{S\_tra}$" means the parasitic resistance measured after connecting the compensation capacitor and the coil in traditional way. Compared by the data in Table 3.7, the parasitic resistance of the coil winding of the

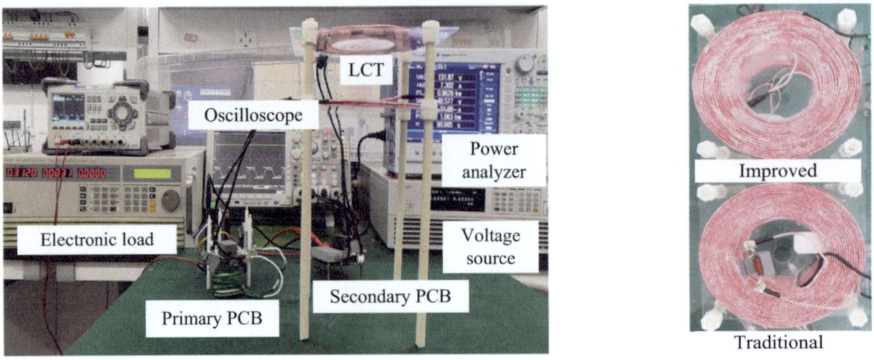

(a) Prototype (b) Coils

Fig. 3.35 IPT prototype with LCC/S compensation and the coils

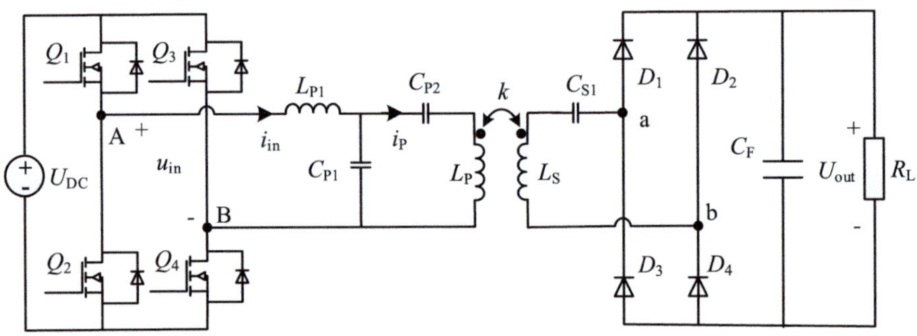

Fig. 3.36 IPT system with LCC/S compensation

Table 3.6 Circuit parameters of the IPT system with LCC/S compensation

Parameter	Value	Unit	Parameter	Value	Unit
f	100	kHz	L_1	9.1	μH
$k_{\_tra}$	0.154		$k_{\_im}$	0.152	
L_{P_tra}	177.5	μH	L_{P_im}	177.2	μH
L_{S_tra}	177.2	μH	L_{S_im}	177.6	μH
C_{P1}	279.9	nF	C_{P2_im2}	28.71	nF
C_{P2_tra}	14.96	nF	C_{S1_im}	14.32	nF
C_{S1_tra}	14.36	nF	C_{S1_im1}	28.82	nF
C_{P2_im}	15.00	nF	C_{P2_im2}	28.47	nF
C_{P2_im1}	31.40	nF	R_L	18–200	Ω

Table 3.7 Parasitic resistance of the LCT and its series compensation capacitor

Parameter	Value (mΩ)	Parameter	Value (mΩ)
R_{LP_tra}	220	R_{LP_im}	212
R_{LS_tra}	220	R_{LS_im}	214
R_{CP2_tra}	66	R_{CP2_im}	67
R_{CS1_tra}	73	R_{CS1_im}	72
$R_{LP\_tra+CP2\_tra}$	286	$R_{LP\_im+CP2\_im}$	246
$R_{LS\_tra+CS1\_tra}$	293	$R_{LS\_im+CS1\_im}$	257

improved method is smaller than that of the traditional method. This is mainly because the coil with the improved winding method has a smaller parasitic capacitance voltage and smaller dielectric loss. On the other hand, in the improved method, the parasitic resistance measured after the coil is compensated is much smaller than the sum of the parasitic resistance of the coil and the compensation capacitor measured under separation conditions. This is because before the coil is compensated in segments, the coil loss mainly includes the dielectric loss of copper loss and parasitic capacitance, and after the coil is compensated in segmentation, the dielectric loss of parasitic capacitance is reduced, and the coil loss is mainly copper loss. The difference between 33 mΩ on the primary side and 29 mΩ on the secondary side is caused by the dielectric loss of the parasitic capacitance. The traditional centralized compensation method does not reduce the loss of parasitic capacitance, so the parasitic resistance of the compensated coil is equal to the sum of the parasitic resistance of the coil and the compensation capacitor measured under separation conditions.

Figure 3.37 shows curve of the output voltage and efficiency of the system varying with load resistance. Compared with the traditional method, the heavy load efficiency of the system using the improved winding and compensation method is increased by 0.74% and the light load efficiency is increased by 2.48%. The increase in system efficiency is even more pronounced at light loads. Experimental results show that the efficiency of IPT system using improved winding and compensation method is higher, because the proposed single-layer segmentation compensation method reduces the dielectric loss of LCT. The equivalent parasitic resistance of the coil is reduced, so the efficiency of the improved method system is higher in the full load range. Figure 3.38 is the waveforms of the circuit under different conditions. The circuit waveforms of the system with traditional and improved winding and compensation methods are basically the same. The input impedance of the system in the full load range is weak inductive, which is conducive to achieving zero voltage switching. The experimental waveforms show that the working state of the prototype using different methods is basically the same, indicating that the improved system and the traditional system are compared under the same conditions.

When the input voltage is 10 V and $R_L = 10\ \Omega$, the voltage difference U_{Cin} between the innermost two-turn coils of the first and second layers of coils and the voltage difference U_{Cout} of the outermost two-turn coils are shown in Fig. 3.39. In the traditional method, the innermost two turns of the first- and second-layer coils

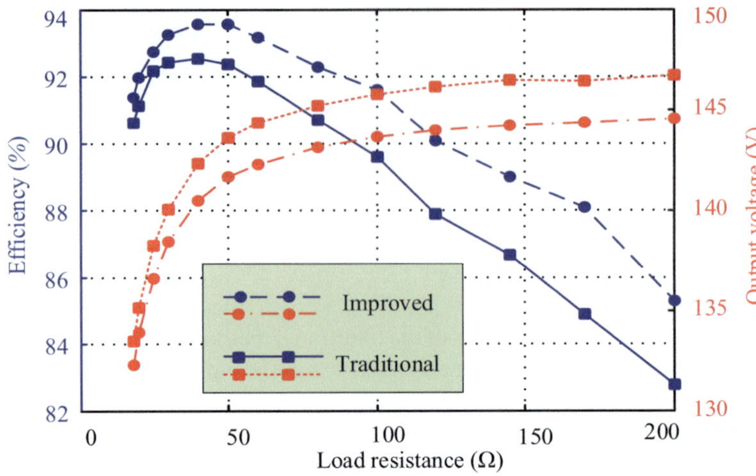

Fig. 3.37 The curves of system output voltage and efficiency varying with load resistance

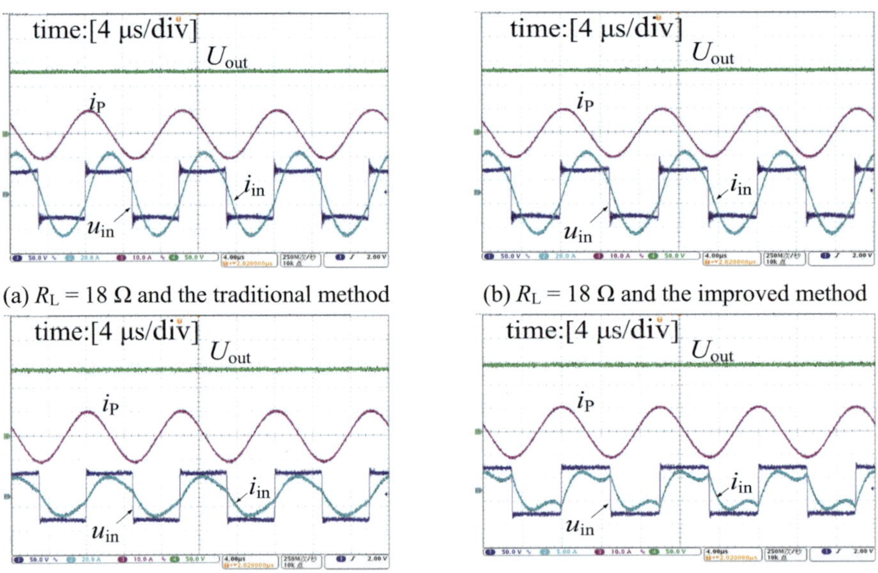

(a) $R_L = 18\ \Omega$ and the traditional method (b) $R_L = 18\ \Omega$ and the improved method

(c) $R_L = 40\ \Omega$ and the improved method (d) $R_L = 200\ \Omega$ and the improved method

Fig. 3.38 Circuit waveforms under different compensation methods and load conditions

are directly connected, so their voltage difference is zero. The voltage difference between the outermost two turns of the coil is 230 V. In the improved method, the voltage difference between the inner and outer turns of the coil is very small, which achieves the design purpose.

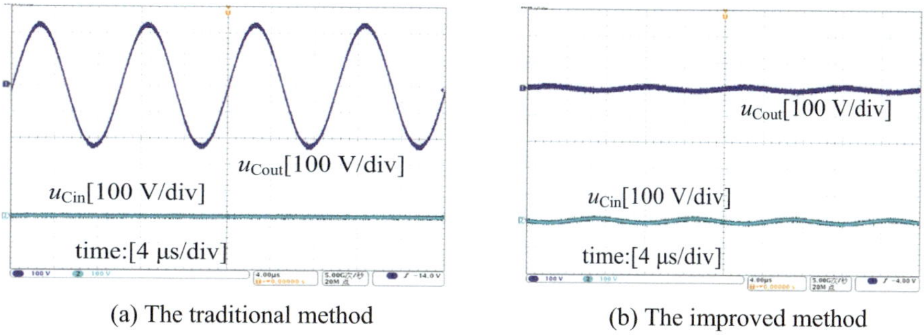

Fig. 3.39 The voltage difference between the innermost and outermost double-turn coils of the coils U_{Cin} and U_{Cout}

3.3 Conclusion

This chapter analyzes the mechanism of parasitic capacitance of loosely coupled transformers, including the following points:

(1) Dielectric losses from parasitic capacitance cannot be ignored. A more accurate coil distribution parameter model is proposed, and a method for calculating coil impedance characteristics and quality factors is given. The effectiveness of the proposed distribution parameter model and calculation method is verified by simulation and experiment.

(2) The single-turn segmented series compensation and single-layer segmented series compensation methods to solve the adverse effects of parasitic capacitance are proposed and reduce the parasitic capacitance loss of tightly wound coils, which is helpful to improve the efficiency of the IPT system. The compensation capacitor is connected in series with sections into the coil, the coil resonates in segments, and the parasitic capacitance is bypassed, which effectively reduces the loss and its effect on the resonant circuit. Single-turn segmented series compensation changes the coil equivalent inductance from capacitive to approximately pure inductance, and the system efficiency of single-layer segmented series compensation is up to 2.48% higher than that of the traditional method. Simulation and experimental results prove the effectiveness of the proposed method.

References

1. Bosshard R, Mühlethaler J, Kolar JW, Stevanović I (2013) Optimized magnetic design for inductive power transfer coils. In: 2013 twenty-eighth annual IEEE applied power electronics conference and exposition (APEC), Long Beach, CA, USA, pp 1812–1819
2. Wang Y, Dongye Z, Zhang H, Zhu C, Lu F (2021) A domino-type load-independent inductive power transfer system with hybrid constant-current and constant-voltage outputs. IEEE Trans Power Electron 36(8):8824–8834

Chapter 4
Analysis and Design on 3D Magnetic Coupling Structure for Long-Range WPT

Abstract This chapter proposes a kind of cubic-magnetic coupler (CMC) for inductive power transfer (IPT) system, which has huge application potential in the charging field of moving targets. The CMC can still maintain sufficient coupling coefficient when used for long-distance wireless power transfer, and it also has great capability of misalignment tolerance. Distance-to-diameter ratio (DDR) is defined to eliminate the impact of coil size and make the comparison between different couplers fairer and easier. This article optimizes the parameters and shape of the CMC. Then, the proposed coupler is compared with several existing couplers employed in long-distance IPT systems. Double-sided LCC compensation topology is chosen to limit the current flowing through the coupling coils. A CMC with an outer size of 180 × 180 × 180 mm^3 was made, based on which an IPT prototype was built. The superiorities of the proposed coupler were verified by a series of experiments. When the DDR equals 3, the coupling coefficient of the CMC is 0.00985, and the maximum dc–dc and ac–ac efficiencies are 64.9% and 72.8%, respectively.

Keywords Distance-to-diameter ratio (DDR) · Double-sided LCC compensation topology · Long-distance inductive power transfer (IPT) · Magnetic coupler · Misalignment tolerance (MT)

4.1 Analysis and Design of Cubic Magnetic Coupler

4.1.1 Distance to Diameter Ratio Index of Wireless Power Transfer

This study uses Eq. (4.1) to reflect the Distance to Diameter Ratio (DDR) of three-dimensional magnetic coupling structures with regular geometric shapes to evaluate the relative distance wireless power transfer between three-dimensional magnetic coupling structures of different shapes and sizes:

Y. Wang et al., *Long Range Inductive Power Transfer*, CPSS Power Electronics Series,
https://doi.org/10.1007/978-981-96-7541-8_4

$$\mathrm{DDR} = \frac{D_{\mathrm{coupler}}}{D_{\mathrm{ED}}} \tag{4.1}$$

where, D_{coupler} is the closest distance between the primary and secondary magnetic coupling mechanisms, and D_{ED} is the equivalent diameter of the coupler. Formula (4.2) is used to determine:

$$D_{\mathrm{ED}} = \mathrm{average}\left\{\frac{V_{\mathrm{primary}}}{S_{\mathrm{min\_primary}}}, \frac{V_{\mathrm{secondary}}}{S_{\mathrm{min\_secondary}}}\right\} \tag{4.2}$$

In Formula (4.2), V_{primary} and $V_{\mathrm{secondary}}$ represent the volume of envelope of the primary and secondary coupling mechanism in 3D space. $S_{\mathrm{min\_primary}}$ and $S_{\mathrm{min\_secondary}}$ represents the minimum projected area on XOY, YOZ, and ZOX surfaces of the primary and secondary coupling mechanism when the bottom surface of the two coils is parallel to the XOY plane, and one side of the bottom surface is parallel to the X-axis. The average equivalent diameter value of the primary coil and secondary coil equals the equivalent diameter of the whole coupler D_{ED}.

In practice, the most widely used three-dimensional magnetic coupling structures are the cube-shaped magnetic coupling structure shown in Fig. 4.1 and the cylindrical-shaped magnetic coupling structure shown in Fig. 4.2. Next, take Fig. 4.1 as an example to better explain this concept. The figure shows a set of cube-shaped loosely coupled transformer, where the parameters of the geometry are marked in the figure. From the above-mentioned concept of D_{ED}, the DDR of the LCT can be calculated. Figure 4.1a shows the 3D structure of two sets of coils. In order to reflect the universal applicability of the DDR concept, each group of loosely coupled transformers consists of two coils, one large and one small. Figure 4.1b is the projection of the magnetic coupling mechanism on the XOZ plane. Due to the symmetry of this structure, the projection of the magnetic coupling structure on the YOZ plane can also be represented by Fig. 4.1b. In addition, Fig. 4.1c shows the projection of the magnetic coupling mechanism on the XOY plane. Wherein, the projection area of the primary side coil and the secondary side coil on the XOZ plane is the smallest, so the equivalent diameters of both are W_{Cub}. Then, Formula (4.2) shows that the equivalent diameters of the whole coupling mechanism are W_{Cub}. Then, the DDR of the loosely coupled transformer can be expressed by Formula (4.3).

$$DDR\_{\mathrm{Cub}} = \frac{D_{\mathrm{coupler\_Cub}}}{D_{\mathrm{ED\_Cub}}} = \frac{D_{\mathrm{Cub}}}{W_{\mathrm{Cub}}} \tag{4.3}$$

In the same way, the DDR of the cylindrical-shaped magnetic coupling structure shown in Fig. 4.2 can be expressed as follows (4.4).

$$DDR\_{\mathrm{Cyl}} = \frac{D_{\mathrm{coupler\_Cyl}}}{D_{\mathrm{ED\_Cyl}}} = \frac{4D_{\mathrm{Cyl}}}{\pi W_{\mathrm{Cyl}}} \tag{4.4}$$

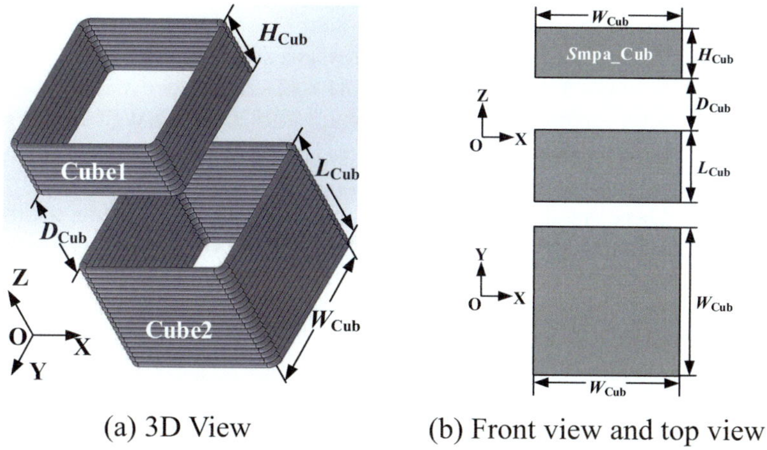

Fig. 4.1 A cubic coupler

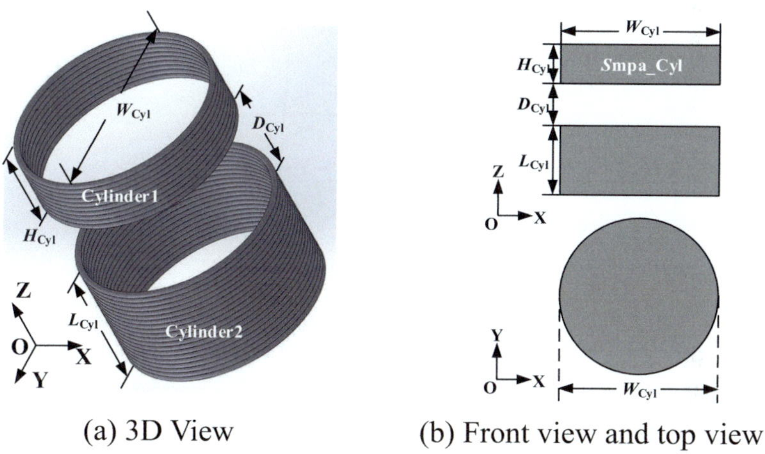

Fig. 4.2 A cylindrical coupler

Using the DDR index, the relative energy transmission distance of the magnetic coupling structure in the WPT system can be better measured.

4.1.2 Design and Optimization of Cubic Magnetic Coupling Structure

Aiming to solve the problems of existing couplers in high DDR IPT system that large in size, low in efficiency, heavy in weight when obtaining same coupling coefficient, this paper referenced the method of Concentrated Magnetic Flux, proposing a LCT which is named Cubic Magnetic Coupler (CMC) as shown in Fig. 4.3.

Figure 4.4 is an exploded schematic diagram of the single-sided magnetic coupling structure in the finite element simulation software ANSYS MAXWELL. The orange part is a Litz coil and the gray part is ferrite core. The entire coil structure is composed of these two parts, both parts are cubic shape.

It is specified here that the height H_{LC} of the coil is not larger than the width W_{LC}, then the minimum projection area of the structure on the XOY YOZ ZOX is the four equal area sides of the Litz coil. According to the definition in (4.2), the equivalent diameter D_{ED} of the coupler is W_{LC}. Therefore, the DDR of the CMC coupler is

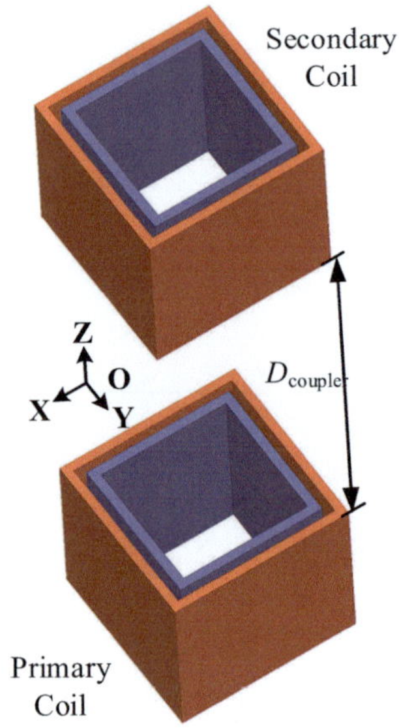

Fig. 4.3 The Cubic Magnetic Coupler

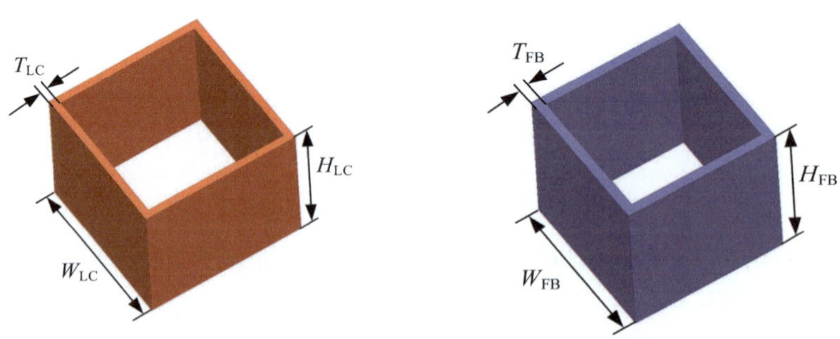

(a) The Litz coil part (b) The ferrite cores part

Fig. 4.4 The exploded pictures of the Cubic Magnetic Coupler

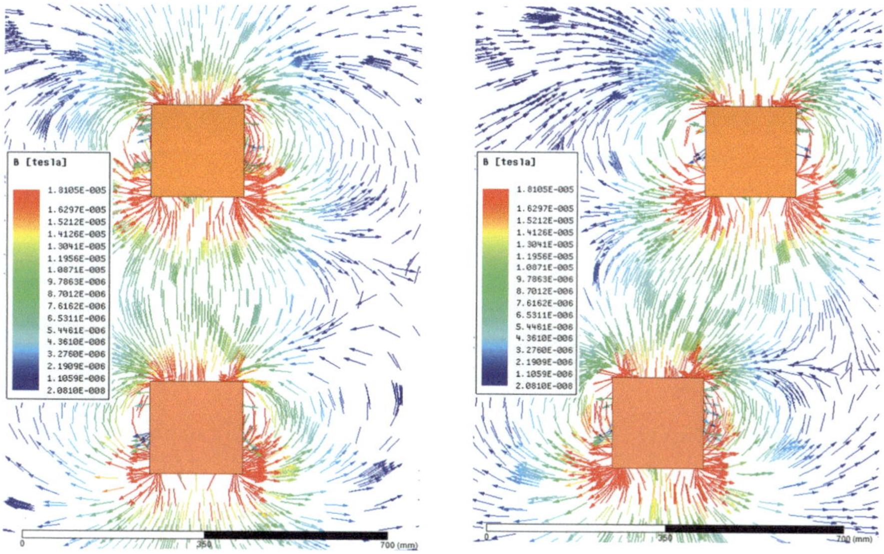

(a) Aligned (b) Greatly misaligned

Fig. 4.5 Space vector magnetic field distribution map of the CMC

$$DDR_{\text{CMC}} = \frac{D_{\text{coupler}}}{W_{\text{LC}}} \tag{4.5}$$

Figure 4.5 is the front view of the space vector magnetic field distribution map of the CMC. Among them, Fig. 4.5a reflects the spatial magnetic field distribution of the entire CMC during power transfer under alignment. Figure 4.5b shows the distribution of the space vector magnetic field when the LCT is misaligned in one direction on the horizontal plane.

It can be seen from Fig. 4.5a that the proposed CMC has a strong magnetic confinement effect on the space vector magnetic field, which effectively enhances the magnetic induction intensity on the IPT path, and can be large under the same floor space. Moreover, it can be seen in Fig. 4.5b that after the LCT has a large deviation in a single direction, the coupling level of the space vector magnetic field is still strong due to the effect of magnetic confinement. It can be considered that the proposed CMC has great misalignment tolerance (MT). Next, fix the primary coil of the proposed CMC unchanged and change the position of the secondary coil in XOY plane to analyze the anti-offset performance of the proposed CMC. The results of the finite element analysis are shown in Fig. 4.6.

In Fig. 4.6, $\delta_{\text{mis-X}}$ and $\delta_{\text{mis-Y}}$ are the Misalignment Rate (MR) of the LCT in the X-axis and Y-axis directions, they represent the relative distance of LCT misalignment. Here, $D_{\text{mis-X}}$ and $D_{\text{mis-Y}}$ are the misalignment lengths of the LCT coil along the X and Y axis, and D_{ED} is the equivalent diameter of the LCT.

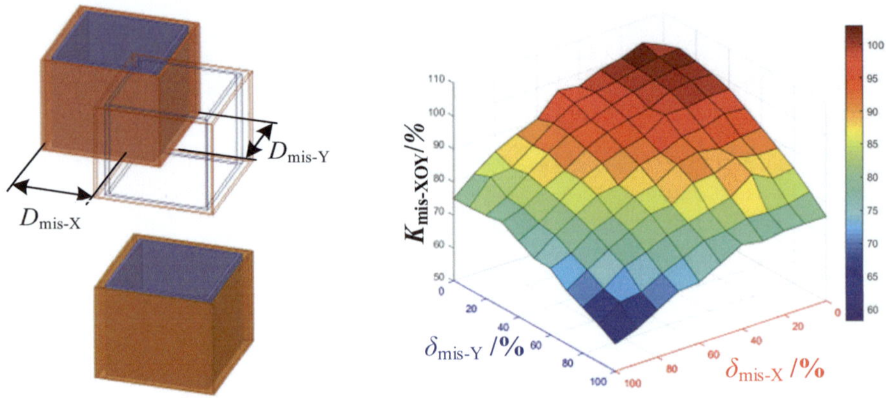

(a) Status of coupler shift (b) Coupling coefficient conditions

Fig. 4.6 Variation trend of the coupling coefficient under misalignment in XOY plane

$$\delta_{\text{mis-X}} = \frac{D_{\text{mis-X}}}{D_{\text{ED}}} \tag{4.6}$$

$$\delta_{\text{mis-Y}} = \frac{D_{\text{mis-Y}}}{D_{\text{ED}}} \tag{4.7}$$

K is the Coupling Coefficient Retaining Percentage (CCRP), reflecting the degree of change in the coupling coefficient before and after the LCT changes, which can be expressed by Formula (4.8). k_{change} is the coupling coefficient after some change of state, and k is the coupling coefficient of the LCT without the above change.

$$K = \frac{k_{\text{change}}}{k} \times 100\% \tag{4.8}$$

It can be seen from Fig. 4.6 that when 2-dimensional misalignment occurs in the XOY plane, the proposed CMC has great capability of MT. Even when the $\delta_{\text{mis-X}}$ and $\delta_{\text{mis-Y}}$ of the coupler is less than 30%, the coupling coefficient of the LCT after the offset is greater than that when it is fully aligned. When the coil is misaligned along X-axis or Y-axis with a δ_{mis} of 100%, the $K_{\text{mis-X/Y}}$ is about 75%. Even if $\delta_{\text{mis-X}}$ and $\delta_{\text{mis-Y}}$ reach 100% at the same time, $K_{\text{mis-XOY}}$ can still maintain the level of 58.3%.

Gu et al. [1] points out that the AC-AC wireless power transfer efficiency of IPT system is determined by Formula (4.9).

$$\eta_{\max} = \frac{k^2 Q_1 Q_2}{(1 + \sqrt{1 + k^2 Q_1 Q_2})^2} \tag{4.9}$$

Among the formula, Q_1 and Q_2 are the quality factors of the primary and secondary side coils of the LCT, which are only related to the coils themselves. In the High DDR system, the value of the coupling coefficient k is usually low, and the quality factor of the primary and secondary side coils of the loosely coupled transformer is at a high level even if the magnetic core is added. Therefore, in the high DDR system, the coupling coefficient k is the determinants of efficiency. In order to maintain the acceptable power transfer efficiency of the system, we expect to obtain a sufficiently large k on the premise of maintaining the large coil quality factors Q_1 and Q_2, so as to meet the requirements of power transfer efficiency.

Next, through the method of multi-dimensional finite element evaluation and analysis, the parameters of the proposed three-dimensional CMC structure are optimized and designed. It is expected to obtain better results in WPT performance and practical applications, and to compare with various existing magnetic coupling structures at the same size.

First of all, it is necessary to be pointed out here that in the field of IPT research, Litz wire diameter T_{LC} and core thickness T_{FB} have little effect on the coupling coefficient. The thickness of ferrite and the stranded wire are usually determined by the power level of the IPT system. In addition, as for the ferrite size W_{FB}, it is also easy to understand that when W_{FB} is similar to W_{LC} relatively, the coupling coefficient will be maximized. Therefore, the influence of these factors is not subjected to excessive optimization analysis.

As shown in Figs. 4.3 and 4.4, the parameters for setting the coupling structure are shown in Table 4.1. In Fig. 4.3, the bottom coil is the transmitter and the top coil is the receiver. Table 4.1 only lists the parameters on one side.

Then, fixing the footprint of the coil W_{LC} to analyze the influence of the coil height H_{LC} and the ferrite core height H_{FB} on the coupling coefficient. The results of finite element simulation analysis are shown in Fig. 4.7. Here, other parameters of the coupler are shown in Table 4.1.

$K_{\Delta\text{H}}$ is the CCRP according to (4.8), which can be seen from Fig. 4.7 that the k of the CMC mainly depends on the H_{FB}. As the height of the magnetic core decreases, $K_{\Delta\text{H}}$ decreases rapidly. Therefore, when designing the CMC, in order to ensure that the loosely coupled transformer can achieve a sufficient k, the H_{FB} should be as large as possible. In addition, H_{LC} also has a certain influence on the coupling coefficient. In the above CMC, when $H_{\text{FB}} = 180$ mm and 90 mm $< H_{\text{LC}} <$ 180 mm, $K_{\Delta\text{H}}$ is slightly greater than 100%. This shows that the H_{LC} concentration in the range of

Table 4.1 Simulation parameters

Parameter	Value	Parameter	Value
f	85 kHz	n	30 mm
I	5 A	H_{LC}	18–180 mm
D_{coupler}	540 mm	W_{FB}	160 mm
T_{LC}	5 mm	T_{FB}	5 mm
W_{LC}	180 mm	H_{FB}	18–180 mm

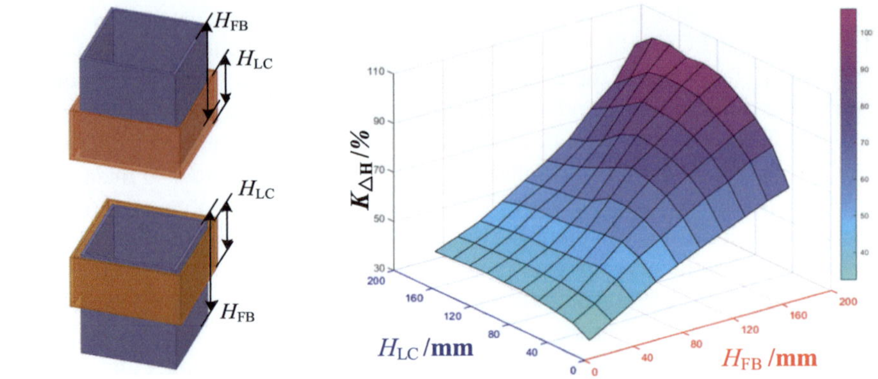

(a) Change of coil parameters (b) Coupling coefficient conditions

Fig. 4.7 Variation trend of the CCRP after changing LCT size

$(0.5H_{FB}, H_{FB})$ can also maintain the k as a relatively high level. Of course, this is for the case where the Litz wire is not completely covered with ferrite, it depends on the power range of the system, the number of the coils' turns and the diameter of the Litz wire.

Next, the dimensions of the fixed magnetic core and Litz coils remain unchanged, and the influence of the position of the Litz coil on the CMC coupling coefficient is analyzed when $H_{LC} = 90$ mm. The simulation results are shown in Fig. 4.8. $K_{\Delta C}$ is the CCRP according to (4.8). From the figure, it is obvious that when the size and position of the ferrite core are fixed, the coupling coefficient increases as the distance between the two Litz coils decreases.

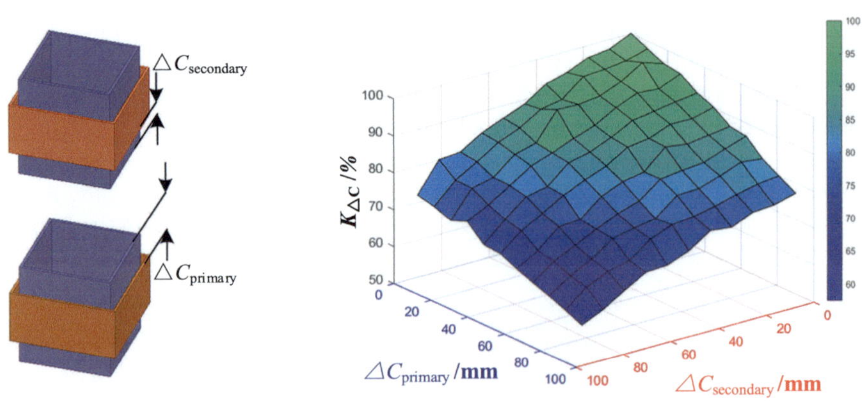

(a) Change of coil positions (b) Coupling coefficient conditions

Fig. 4.8 Variation trend of the CCRP after changing the coils' position

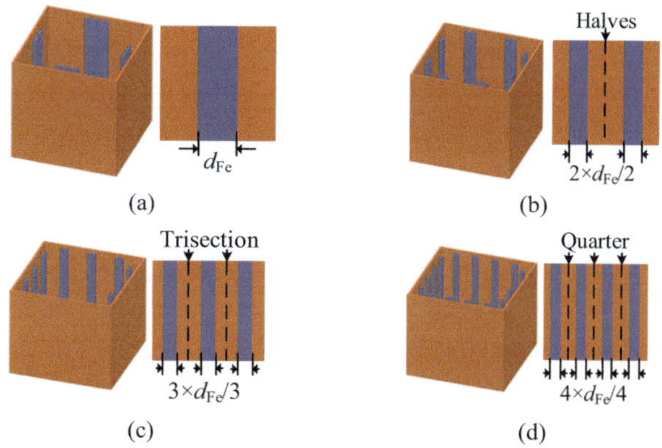

Fig. 4.9 Conditions of the CMC's ferrite core in different longitudinal split states and 2D diagrams on each of its inner surfaces

As the proposed CMC uses a high proportion of ferrite cores, this makes the weight and cost of the LCT at a higher level. From the magnetic field distribution of LCT in Fig. 4.5, it can be seen that the ferrite core plays an important role in the increase of magnetic induction, and the magnetic induction is obviously restricted and enhanced in the longitudinal direction. However, this magnetic field distribution is achieved when the core is completely covered in the Litz coil. If the coupling coefficient of the LCT can be made little change while reducing the laying density of the ferrite core, the weight and cost problems of the structure can be effectively solved. Next, this study proposes a method of Longitudinal Splitting of the Ferrite Core (LSFC) to optimize the magnetic core structure to reduce the weight and cost of the coupler.

As shown in Fig. 4.9a–d, the magnetic core with length d_{Fe} on each side is divided into 1–4 equal parts longitudinally, and the geometric centerlines of multiple magnetic strips on each side are uniform distributed inside the Litz coil. d_{Fe} varies from 10 to 140 mm. In addition, the 2D pictures in each case represent the arrangement of one wall inside a CMC.

Under the four subdivisions, the variation trend of the LCT coupling coefficient is shown in Fig. 4.10. Here, DDR of the proposed CMC is 2, H_{LC} and H_{FB} are both 180mm, and other parameters of LCT are shown in Table 4.1. K_{Fe} is the CCRP when the ferrite core is changed according to (4.8), it reflects the change of the coupling coefficient after the distribution of the ferrite core changes. What needs to be explained here is that when K_{FE} is 100%, the four surfaces and four corners are filled with ferrite cores as Fig. 4.3. In the curves of the four cases in Fig. 4.10, when the core length reaches 150 mm, the four volumes are There is no magnetic core in 4 corners of $5 \times 5 \times 180$ mm^3, which is the main reason why $K_{FE} < 100\%$ when $d_{FE} = 150$ mm in the four cases.

It can be seen from the trend in Fig. 4.10 that K_{Fe} and d_{Fe} show a positive correlation. When $K_{Fe} < 95\%$, under the same d_{Fe} condition, the more parts the ferrite core is split by LSFC, the larger the K_{Fe} is. Even when the magnetic core is divided into four equal parts, when $d_{Fe} = 70$ mm, K_{Fe} is still about 90%, that is, when the amount

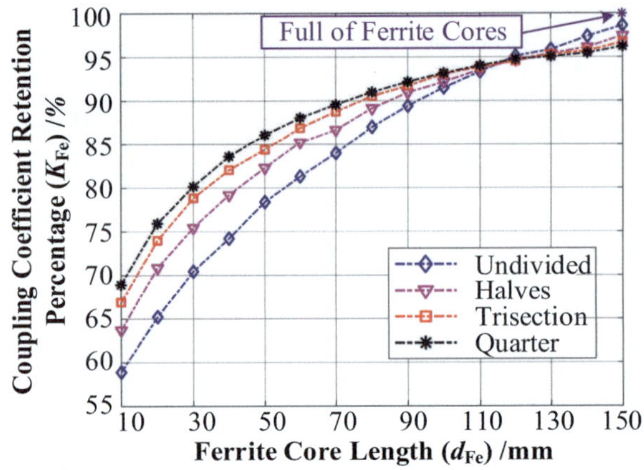

Fig. 4.10 CCRP in different states of ferrite core

of magnetic core becomes less than half, the coupling coefficient is only reduced by 10%. In this way, a practical and applicable conclusion could be obtained that the method of LSFC can effectively reduce the weight and cost of LCT. However, obviously, after the above changes, there is always $K_{Fe} < 100\%$, which shows that the coupling coefficient of the proposed CMC is the largest when the inner layer is covered with the core as shown in Fig. 4.3.

The following is a summary of the design and optimization of the above 3D CMC structure. First of all, the ferrite core is the determining factor of the CMC's coupling coefficient. In the actual design, the H_{FB} should be increased as much as possible. At the same time, the method of LSFC can be used to optimize the distribution of the magnetic core and achieve the purpose of reducing the weight and cost of the CMC, the optimization solution is conducive to the commercial application of the CMC. At the same time, the height H_{LC} and position ΔC of the Litz coil also have a certain influence on the coupling coefficient. In the actual design, actual factors such as wire thickness, turns and system power can be combined to optimize the distribution of Litz coils.

4.1.3 Design and Optimization of Cubic Magnetic Coupling Structure

Next, the proposed CMC is compared with 6 other LCTs, focusing on the variation in the coupling coefficient of different LCTs under different DDR values. Moreover, comparing their misalignment tolerance (MT) under typical DDR values.

First, we number the 7 LCTs and unify their sizes, so that the comparisons under similar sizes can ensure fairness to a certain extent, according to (4.8), $D_{ED, \#1\text{–}\#7} = 180$ mm. The structure and size of the other 6 LCTs are shown in Fig. 4.11.

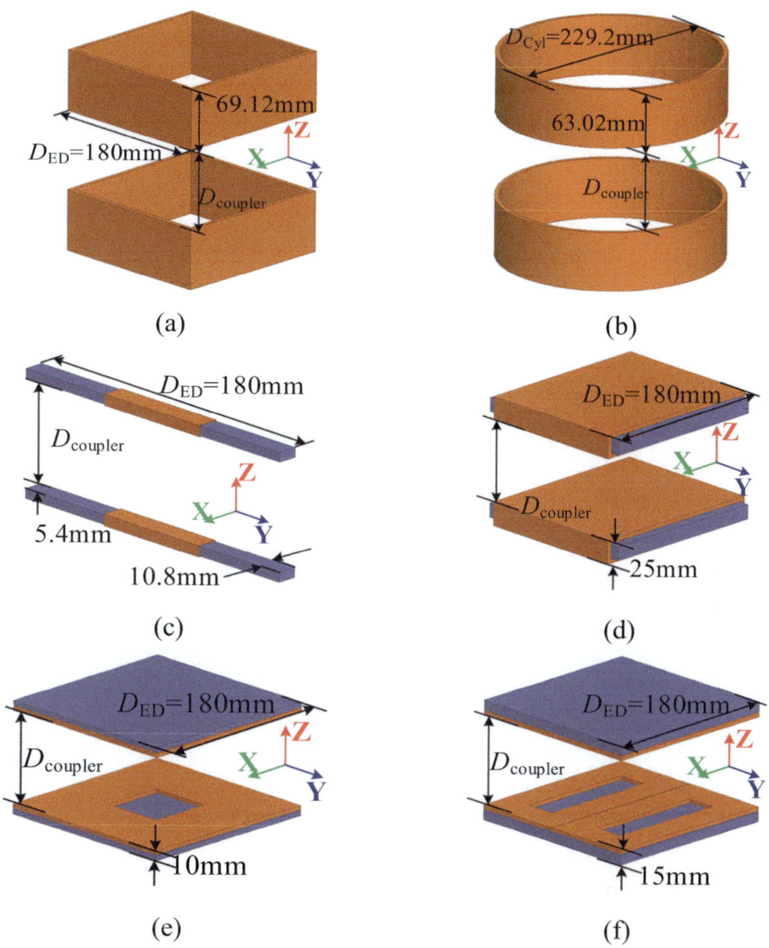

Fig. 4.11 Schematic diagram of different LCTs after uniform size

#1: The CMC. Its structure is shown in Fig. 4.3, and the parameters are shown in Table 4.1, H_{LC} and H_{FB} are both 180 mm. $D_{ED} = W_{LC} = 180$ mm.

#2: The Square Solenoid Coils proposed by Deng et al. [2]. Setting $D_{ED} = 180$ mm to make its size uniform with #1, and adjusting the other parameters according to the proportion. The details are shown in Fig. 4.11a.

#3: The Helix Shape Coils proposed by Zhou et al. [3]. Setting $D_{ED} = 180$ mm, the cylinder diameter $D_{cyl} = 229.2$ mm which is calculated by (4.2). Adjusting the other parameters according to the proportion. The details are shown in Fig. 4.11b.

#4: Dipole-coil proposed by Choi et al. [4]. Set $D_{ED} = 180$ mm to make its size uniform with #1, and adjust other parameters according to the proportions. The details are shown in Fig. 4.11c.

#5: Classic flat solenoid LCT. The details are shown in Fig. 4.11d.

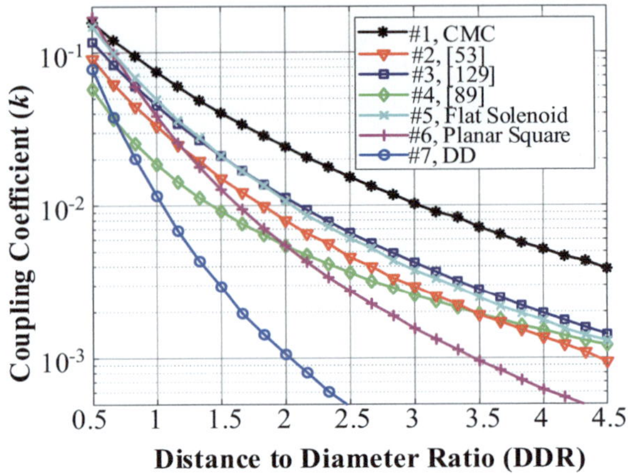

Fig. 4.12 Variation of coupling coefficient of 7 LCTs with uniform size under different DDR

#6: Classic planar square LCT. Both two coils come with ferrite cores, and the details are shown in Fig. 4.11e.

#7: Classic DD shape LCT. Ferrite cores are attached to the back of both coils, and the details are shown in Fig. 4.11f.

By changing the above 7 LCTs' D_{coupler}, the influence of DDR for different couplers on the coupling coefficient is analyzed. The DDR of the 7 LCTs was changed between 0.5 and 5, by using finite element analysis software, the trend could be obtained as shown in Fig. 4.12. Among them, due to the large change scale of the coupling coefficients, the Y-axis adopts the log counting method.

After the above comparative analysis, it is clearly that the k of CMC is significantly larger than other LCTs almost when DDR > 1, that provides insurance of enough efficiency for high DDR IPT system. Although #2, #3, #4 are also 3D LCTs, when the DDRs increase, their coupling coefficients decrease steeply, the variation curves of their coupling coefficients are similar to that of #5 flat solenoid LCT. When DDR > 2, the coupling coefficients of #6 and #7 are the worst.

Next, comparing and analyzing the MT of the 7 LCTs under typical DDRs. In the XOY plane, the MT of #4, #5 and #7 in the X axis direction is stronger, therefore, the X-axis is selected as the misalignment direction. Figure 4.13a–c are the relationship between MR $\delta_{\text{mis-X}}$ and CCRP $K_{\text{mis-X}}$ when the DDR values are 0.5, 1 and 2 respectively.

It can be found that the trend curves of 7 LCTs are similar under different DDRs. The proposed #1 and flat solenoid #5 have great capability of misalignment tolerance (MT), #2, #3, #7 also have a certain MT, #4 and #6 are the worst.

Here, the MT of the CMC is similar to #5, but according to Covic and Boys [5], the MT of #5 in Y-axis direction is seriously weaker than the X-axis performance, and the analysis of Fig. 4.6 shows that #1 has great performance in both X and Y axis. Therefore, the CMC has the best comprehensive ability among 7 LCTs in terms of MT.

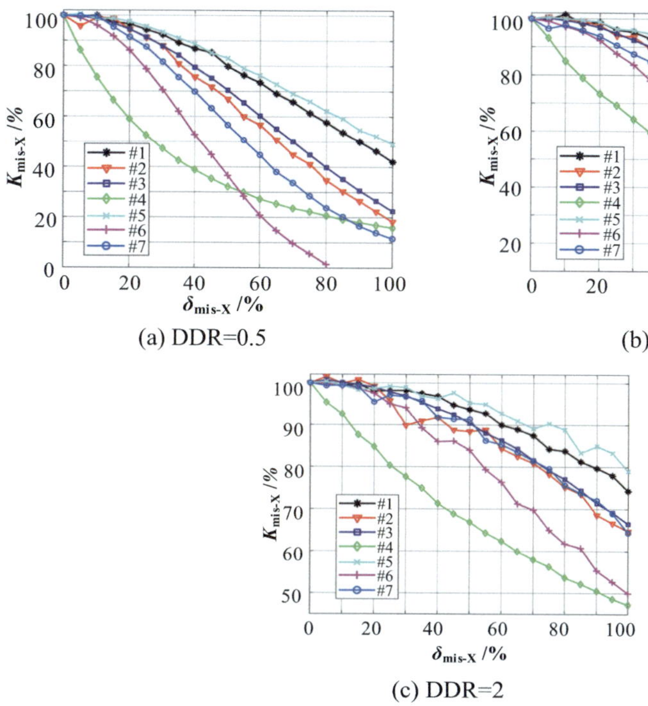

Fig. 4.13 Variation of $K_{mis\text{-}X}$ of 7 LCTs with different $\delta_{mis\text{-}X}$ under three typical DDRs values

4.2 Long Air Gap WPT System Based on Three Dimensional Cubic Magnetic Coupler

4.2.1 Establishment of Long Air Gap Wireless Power Transfer System

The conventional IPT system circuit consists of a DC power supply, a full-bridge inverter, a loosely coupled transformer, a compensation topology, a rectifier bridge, a filter capacitor, and a load. Next, the proposed CMC in Sect. 4.2 is placed in the IPT system in Fig. 4.14 to analyze the electrical characteristics of the AC side of the entire system.

Double-sided LCC compensation topology was proposed by Chris Mi, a well-known scholar in the field of WPT research. This structure can control the electrical quantity on the AC side of the system by adjusting the primary and secondary compensation inductors L_{f1} and L_{f2}, making the circuit design of the IPT system higher degrees of freedom.

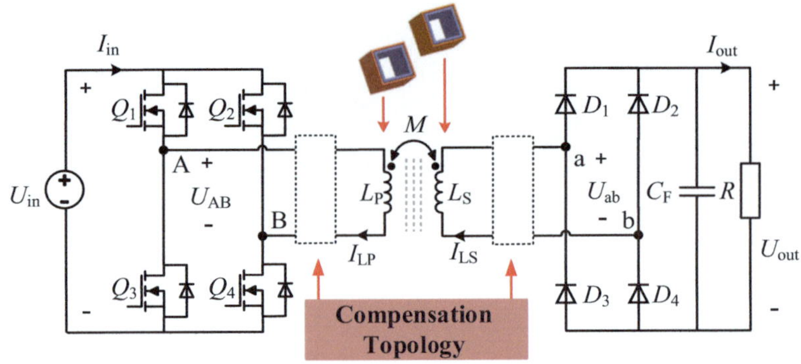

Fig. 4.14 The long air gap IPT system with CMC

The equivalent circuit model of the IPT system based on the double-sided LCC compensation topology is shown in Fig. 4.15. Assume that the inductor and capacitor are ideal components and ignore their parasitic resistance. Among them, L_P and L_S are the self-inductance of the primary and secondary coils, C_P and C_S are the compensation capacitors connected in series with the coil in the circuit, L_{f1}, L_{f2} and C_{f1}, C_{f2} are the inductance and capacitance of the LC resonance network, and R_{eq} is the inductance AC equivalent resistance. The Kirchhoff equations of the circuit can be used to calculate the current expression of each loop of the system:

$$I_{AB\_LCC} = -\frac{M^2 R_{eq} U_{AB\_LCC}}{\omega^2 L_{f1}^2 L_{f2}^2} \tag{4.10}$$

$$I_{LP} = -j\frac{U_{AB\_LCC}}{\omega L_{f1}} \tag{4.11}$$

$$I_{LS} = \frac{U_{AB\_LCC} M R_{eq}}{\omega^2 L_{f1} L_{f2}^2} \tag{4.12}$$

$$I_{ab\_LCC} = -j\frac{M U_{AB\_LCC}}{\omega L_{f1} L_{f2}} \tag{4.13}$$

It can be known from Formulas (4.10) to (4.13) that given the input voltage and load, the inductance of the compensation inductors L_{f1} and L_{f2} can be adjusted to effectively control the current through the primary and secondary coils and the input and output current of the AC system. This can make a high DDR IPT system run efficiently without high power loss due to excessive branch current.

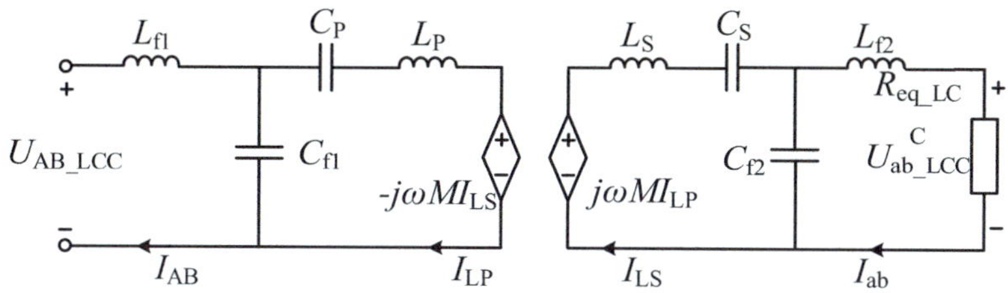

Fig. 4.15 AC-side equivalent circuit model of IPT system based on double-sided LCC compensation topology

4.2.2 Experimental Verification

To verify the excellent performance of the High DDR system based on the proposed coupler, a CMC with help of 3D printing technology has been designed, and the secondary coil of the coupler as shown in Fig. 4.16. Where, the primary side coil is 30 turns, the secondary side coil is 45 turns, H_{LC} and H_{FB} are both 180 mm, W_{FB} is 154 mm, and other parameters are shown in Table 4.1. In addition, because the bending of the Litz wire requires a certain arc, the corner of the CMC is designed to be rounded.

In order to prove the proposed CMC has great MT, fixing DDR of the LCT as 2, setting the position of the secondary coil unchanged, and moving the primary coil along the X-axis to achieve misalignment. $\delta_{mis\text{-}X}$ changes from 0 to 100%, measured k and calculated CCRP of the CMC are compared with the simulation data, the result is shown in Fig. 4.17.

The curves in Fig. 4.17 show that the designed LCT has great capability of misalignment tolerance. Among them, the X-axis MR changes from 0 to 100%. In the red broken line, k of the LCT changes from 0.0213 to 0.0159, which is slightly

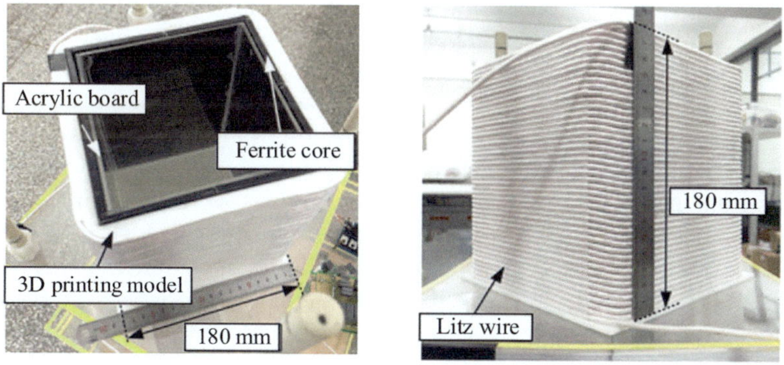

(a) Top view. (b) Side view.

Fig. 4.16 Secondary side of the CMC

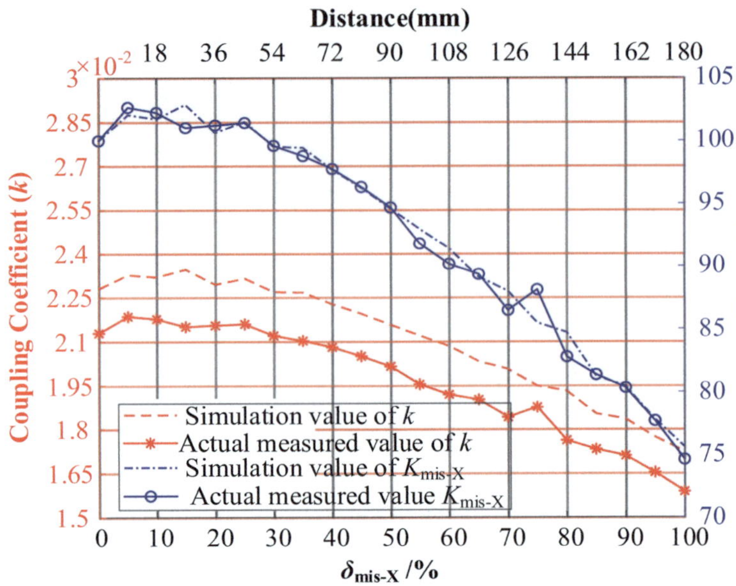

Fig. 4.17 The experimental results of the CMC misalignment tolerance along X-axis direction

different from the simulation results. This is caused by actual factors such as the coupler's magnetic flux leakage and Litz coil fillet in the actual design process. In the blue broken line, the retention rate of the CCRP changes from 100 to 74.6%, which basically coincides with the simulation result of the blue dotted line. It can be seen that the designed coupler conforms to the aforementioned theoretical analysis results.

A prototype experiment platform as shown in Fig. 4.18 based on the coupler was built. The prototype refers to the circuit structure shown in Fig. 4.14 and adopts a double-sided LCC compensation topology. In the experiment platform, parts of the numbers from 1 to 11, respectively are: Oscilloscope, DC power supply, The full bridge inverter circuit, Primary side LCC (L_{f1}, C_{f1}, C_P) compensation circuit, Primary side coil of the CMC, Secondary side coil of the CMC, Secondary side LCC (L_{f2}, C_{f2}, C_S) compensation circuit, Bridge rectifier circuit, Load, 1Non-metallic mechanical platform (acrylic, nylon, wood), Power analyzer.

The actual parameters and measured ESR values of circuit components of Figs. 4.14 and 4.15 are shown in Table 4.2. Where, the measured k is 0.00985 when the distance of two coils is 54 cm, which is similar to the simulation result in Fig. 4.12. In addition, due to practical factors, there is a slight difference between the actual measured value and the theoretical value of the double-sided LCC compensation topology.

First, changing the position of the secondary side coil, the power transfer characteristics of the wireless power transfer system based on the CMC under the condition of changing DDR are investigated by. Setting the load to 30 Ω, maintain the output

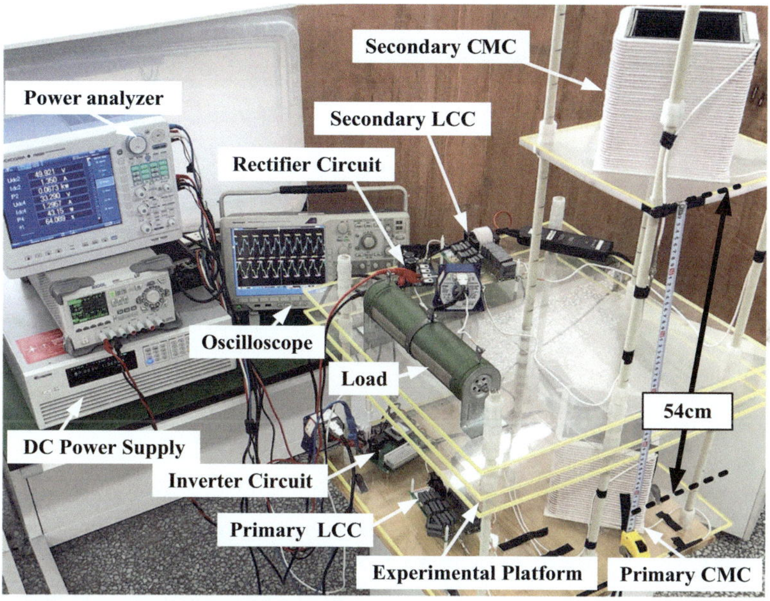

Fig. 4.18 Prototype of the system

Table 4.2 The measured parameters of the system

Parameter	Value/type	Parameter	Value
f_0	85 kHz	C_{f1}	217.92 nF
Q_1–Q_4	IRFBA22N50	R_{Cf1}	9.2 mΩ
D_1–D_4	MBR40200WT	C_P	9.51 nF
L_P	383.9 μH	R_{CP}	85 mΩ
R_{LP}	176 mΩ	L_{f2}	17.03 μH
L_S	867.2 μH	R_{Lf2}	66 mΩ
R_{LS}	318 mΩ	C_{f2}	204.4 nF
$k_{D=54\ \text{cm}}$	0.00985	R_{Cf2}	14 mΩ
L_{f1}	15.94 μH	C_S	4.12 nF
R_{Lf1}	64 mΩ	R_{CS}	163 mΩ

power around 30 W is changeless, by adjusting the input voltage, to change the input power, size, and then examine system of DC-DC efficiency, the experimental results are shown in Fig. 4.19.

Allowing the distance between two coils of the CMC change from 46.5 to 81 cm, that is, DDR change from 2.58 to 4.5. DC-DC efficiency decreases from 68.4 to 26.3%, and with the increase of DDR, the power transfer efficiency of the system decreases at a faster rate. When DDR is less than 4, the system efficiency is above 40%; when DDR is less than 3, the system efficiency is above 60%. It can be seen that

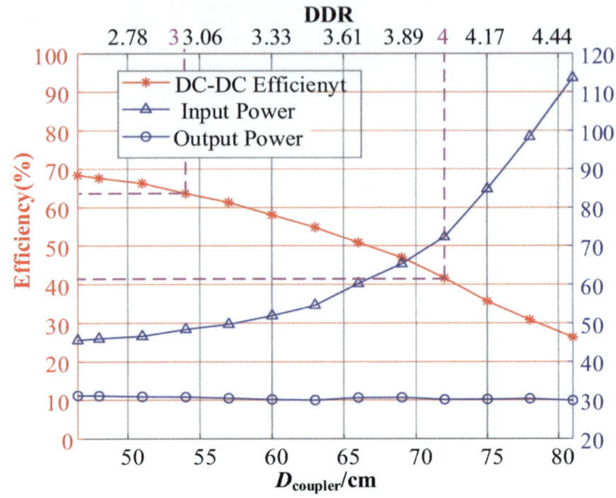

Fig. 4.19 Curve of DC-DC efficiency at a given output power of the system with different DDR

the proposed CMC has excellent High DDR power transfer characteristics. Fixing DDR of the CMC as 3, the distance between two coils is 54 cm, as shown in Fig. 4.18. With the input voltage unchanged at 50 V, the double-sided LCC compensation topology was investigated by changing the resistance value of load *R*, so as to investigate the characteristics of the double-sided LCC compensation topology when changing the load constant voltage input constant current output, and analyze the change of DC-DC efficiency in the process of changing *R*, the curves are shown in Fig. 4.20.

Among them, the red dashed line is the theoretical calculation result obtained by the main circuit parameters and ESR in Table 4.2, the red and blue broken lines represent the experimental measured values of DC-DC efficiency and output current, respectively. As can be seen from Fig. 4.20, with the increase of the load *R* value, the output current decreases approximately linearly, and its decreasing speed is relatively slow. *R* increases in a wide range from 8.8 to 51.3 Ω, I_{out} only drops from 1.45 to

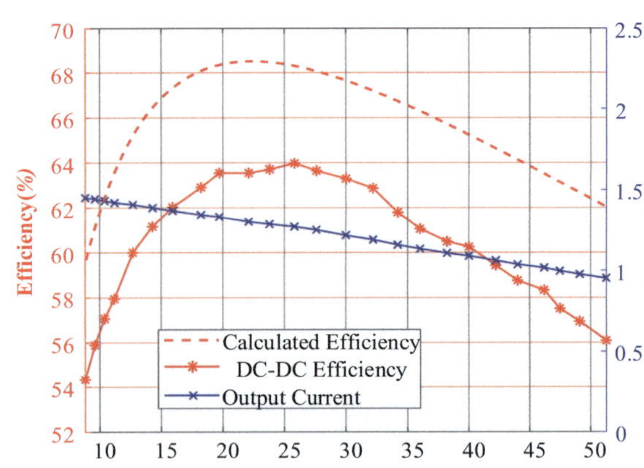

Fig. 4.20 Curve of DC-DC efficiency and output current at a constant input voltage with different load

0.95 A. It can be considered that the high DDR IPT system based on the CMC has good constant current output characteristics. What needs to be pointed out here is that the calculated efficiency and the trend of DC-DC efficiency are similar, but there are some differences in the values, which are caused by additional hysteresis loss, eddy current loss, connection line loss and circuit modeling deviations, etc.

In addition, when the load is 25.8 Ω, the DC-DC efficiency of the system reaches an extreme value of 63.98%, which is the optimal load of the system under this state. The voltage and current waveform are shown in Fig. 4.21. It can be seen that in Fig. 4.21a, the AC input voltage of the system is slightly ahead of the AC input current of the system, the AC system is in a state of resonance, and the switch tube realizes soft switch. AC current waveform is distorted by the double-sided LCC compensation topology. Due to the existence of harmonics, the waveform of DC voltage and current in Fig. 4.21b fluctuates to a certain extent, which will bring harmonic pollution to the power grid. In the practical application of the system, harmonic interference can be eliminated by adding filters. According to the waveform of DC voltage and current, the input and output voltage is about 50 and 33 V, and the current is about 1.4 and 1.3 A. At this point, the quantitative analysis of the system state is shown in Fig. 4.22.

According to the quantitative test by YOKOGAWA PX8000 power analyzer, the maximum efficiency of DC-DC and AC-AC in the above state is 64.9% and 72.8% respectively. Currently, the input and output power of the DC side are 66.1 W and 42.9 W respectively, and the input and output power of the AC side are 62.4 W and 45.4 W respectively. Among them, the loss power of the primary PCB is 3.7 W, that of the secondary PCB is 2.5 W, the total loss power of the main circuit including the compensation topology and the CMC is 17.0 W.

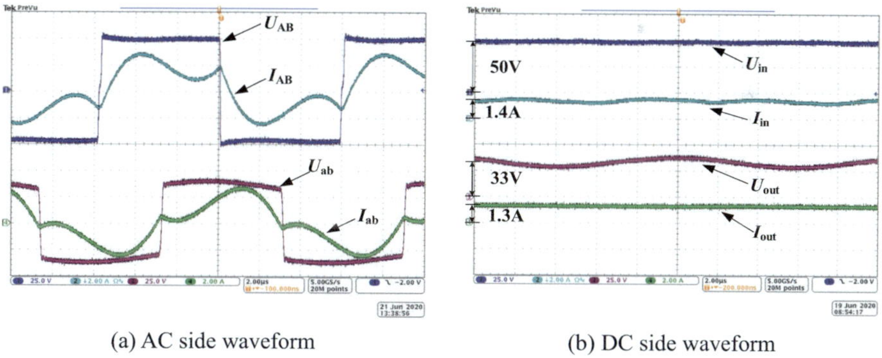

(a) AC side waveform (b) DC side waveform

Fig. 4.21 Waveform of voltage and current when the input voltage of 50 V, load 25.8 Ω and DDR is 3

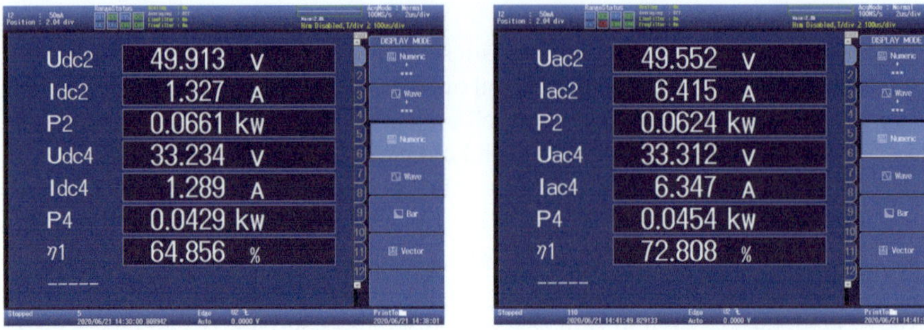

(a) DC-DC efficiency (b) AC-AC efficiency

Fig. 4.22 The power situation of the AC side and the DC side of the system at the maximum efficiency point

4.3 Design and Analysis of Improved Cylindrical Solenoid Coupler

Corresponding to the planar circular and planar square magnetic structures in the planar magnetic coupling structure, cylindrical and cubic shapes are also the basic types in the 3D magnetic coupling structure. Therefore, the above analysis and design methods for 3D magnetic cubic coupling structures are also applicable to cylindrical shapes.

This section proposes an improved 3D solenoid magnetic coupling structure (Cylindrical Solenoid Coupler, CSC) based on the analysis and design method of the three-dimensional magnetic cubic coupling structure, combined with the magnetic field distribution characteristics of the hollow three-dimensional solenoid and the planar solenoid.

The single-stage 3D solenoid is shown in Fig. 4.23. It consists of a Litz coil part and a ferrite core part. Its parameters are set as shown in Table 4.3. Here, L_F, T_F and H_F are the length, thickness and height of each ferrite core unit after the core is longitudinally divided, and θ_F is the spacing angle after the core is longitudinally divided. In addition, D_C, H_C and T_C are the diameter, height, and thickness of the Litz coil respectively. Among them, the larger the coil diameter D_C is, the greater the coupling coefficient of the two-stage magnetic structure is obviously. Therefore, fix the outer diameter of the coil to 230 mm, place the ferrite core close to the coil, and set the D_F to 210 mm. In addition, according to the actual size of the Litz wire and ferrite core material, both T_C and T_F are set to 5 mm.

Usually, the 3D solenoid structure is shown in Fig. 4.24a. Among them, the yellow part is the Litz coil and the gray part is the ferrite core. However, the structure of the inner layer covering the ferrite core increases the weight and cost of the coil, and the curved core with appropriate curvature needs to be specially customized to achieve the goal of full coverage, which further increases the difficulty of designing the magnetic coupling structure. Therefore, this study uses the core longitudinal

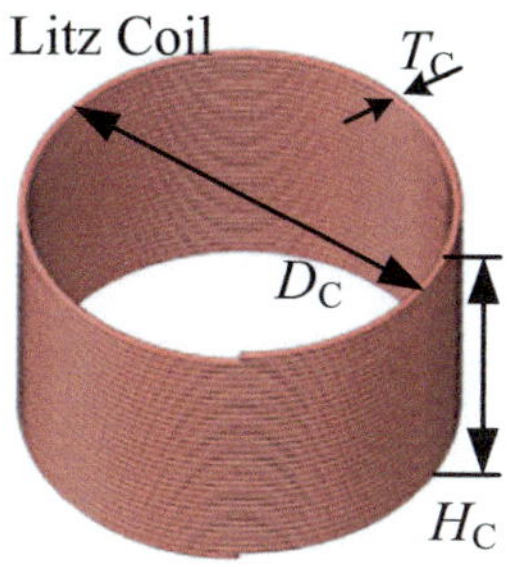

(a) The Litz coil part

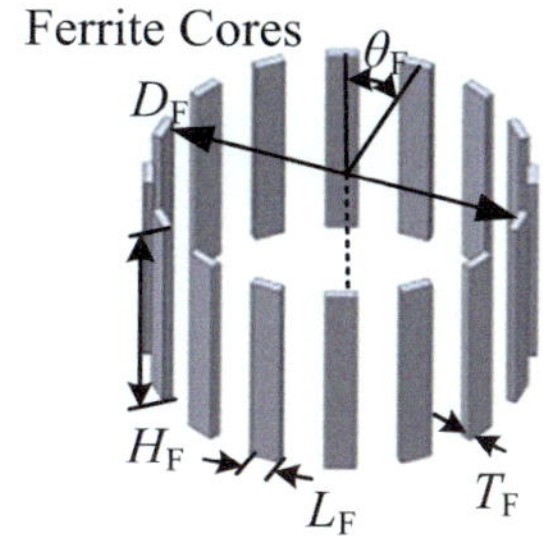

(b) The ferrite cores part

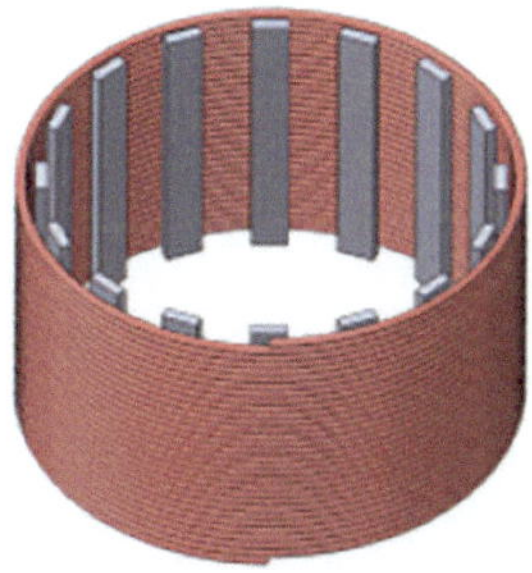

(c) A single coupler

Fig. 4.23 The improved 3D solenoid magnetic coupling structure

Table 4.3 Simulation parameters

Parameter	Value (mm)	Parameter	Value
D_C	230	T_F	5 mm
T_C	5	H_F	10–180 mm
H_C	10–180	L_F	15 mm
n	30	θ_F	15° 22.5° 45°
D_F	210		

splitting method to optimize the structure of the ferrite core to reduce the weight and cost of the coupler. Divide the ferrite core with a total length of d_{Fe} into 8, 16 and 24 equal parts and insert it into the inner layer of the Litz coil, as shown in Fig. 4.24b–d.

Taking the two-stage coupling structure as an example, the trend of the coupling coefficient changing with d_{Fe} at different times of subdivision is shown in Fig. 4.25. Among them, K is the coupling coefficient retention rate.

Through the finite element simulation method, the coupling coefficient variation trend is obtained as shown in Fig. 4.25. Among them, the full magnetic core situation in Fig. 4.24a corresponds to $K_{Fe} = 100\%$ in Fig. 4.25. It can be seen from the trend in Fig. 4.25 that K_{Fe} and d_{Fe} are positively correlated. Under the same d_{Fe} conditions and $K_{Fe} < 99\%$, the more times the ferrite core is longitudinally split, the greater the K_{Fe}. In addition, when the core is divided 24 times, K_{Fe} is still greater than 95% even

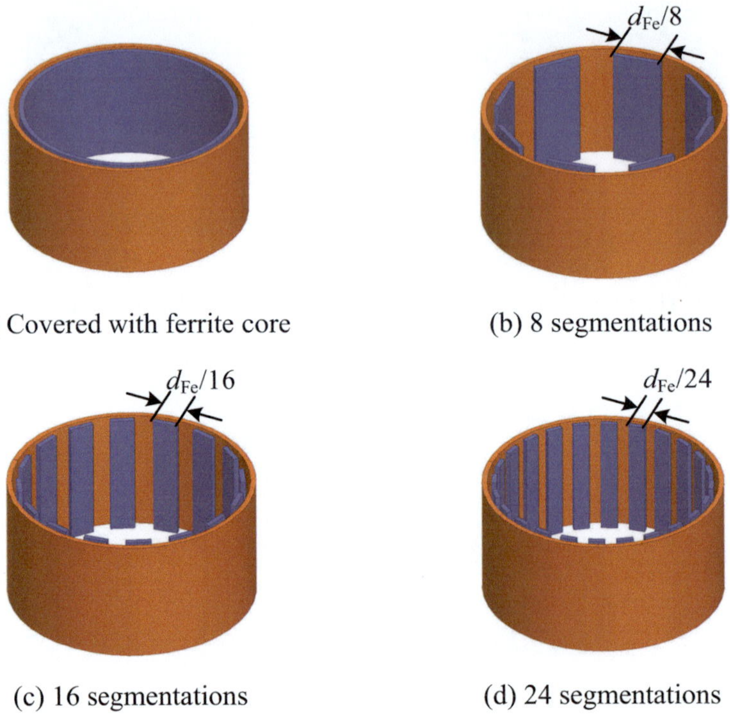

Fig. 4.24 Single coupling structure diagram optimized by the method of Longitudinal Splitting of the Ferrite Core (LSFC)

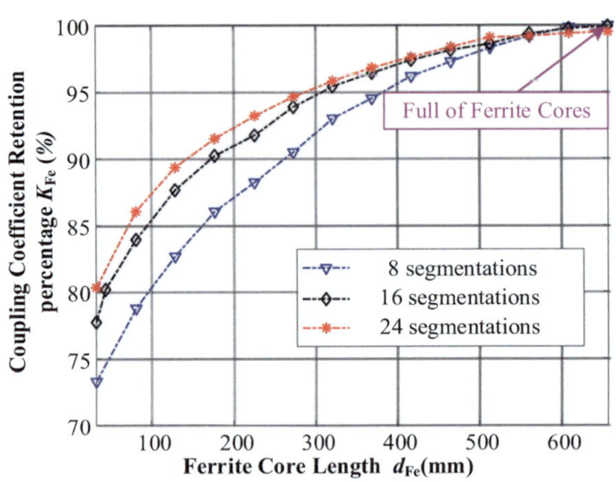

Fig. 4.25 CCRP under different states of ferrite core

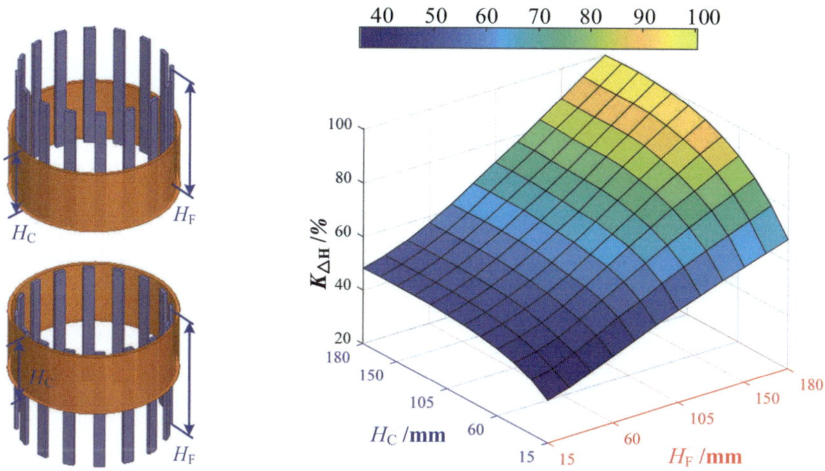

(a) Change of coil parameters (b) CCRP after changing coupler size

Fig. 4.26 Variation trend of the CCRP after changing coupler size

when $d_{Fe} = 321.6$ mm. That is, when the amount of magnetic core becomes less than half of Fig. 4.24a, the coupling coefficient is only reduced by 5% compared with the case of full magnetic core. Therefore, a practical conclusion can be drawn, that is, the weight and cost of the magnetic coupling structure can be effectively reduced by using the optimization method of longitudinal splitting of the magnetic core.

Next, based on the finite element analysis method, a two-stage three-dimensional solenoid coupling structure model was used to analyze the influence of the coil height H_C and the ferrite core height H_F on the coupling coefficient. The finite element simulation analysis results are shown in Fig. 4.26. Here, other parameters of the coupler are shown in Table 4.3.

As shown in Fig. 4.26, $K_{\Delta H}$ is the coupling coefficient retention rate of the variable magnetic coupling structure size. It can be seen that the coupling coefficient of the magnetic structure mainly depends on H_F. As the core height decreases, $K_{\Delta H}$ decreases rapidly. Therefore, when designing the magnetic structure, in order to ensure that the magnetic coupling structure can achieve a sufficient coupling coefficient, the H_F should be as large as possible. In addition, H_C also has a certain impact on the coupling coefficient, but its effect is weak.

Next, with the sizes of the fixed magnetic core and Litz coil unchanged, the influence of the position of the Litz coil on the coupling coefficient of the magnetic structure was analyzed when $H_C = 60$ mm. The simulation results are shown in Fig. 4.27. Among them, $K_{\Delta C}$ is the coupling coefficient retention rate of the variable Litz coil position. According to the results in the figure, when the size and position of the ferrite core are fixed, the coupling coefficient between the two Litz coils increases as the distance decreases.

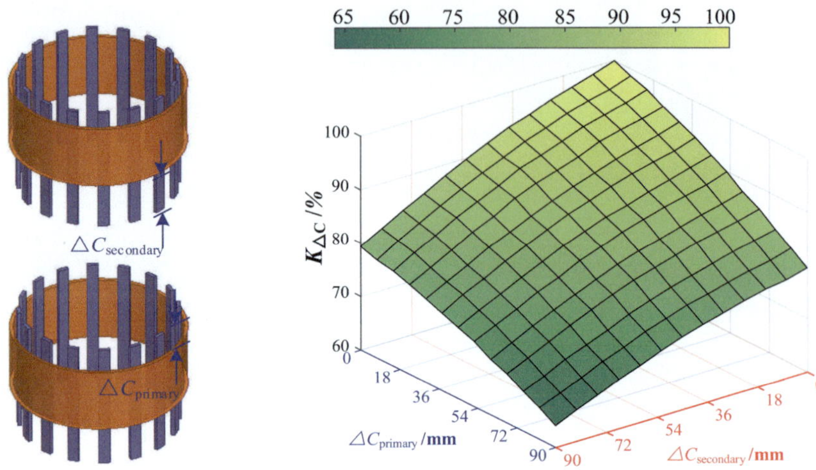

(a) Change of coils' position (b) CCRP after changing coils' position

Fig. 4.27 Variation trend of the CCRP after changing the Litz coils' position

The CSC proposed in this article is improved from the hollow three-dimensional solenoid and planar solenoid shown in Fig. 4.28. In order to verify that this magnetic structure can effectively enhance the coupling level of the magnetic coupling structure under the long air gap, the 3D magnetic cubic coupling structure with a size of 180 mm * 180 mm * 180 mm proposed in Sects. 4.2 and 4.3 was combined. Through finite element analysis, the trend of the coupling coefficients of the four coupling structures changing with the WPT distance under the same size is obtained, as shown in Fig. 4.29. The proposed CSC parameters are shown in Table 4.3, $H_C = H_F =$ 180 mm, $\theta_F = 22.5°$. According to Formula (4.2), the equivalent diameter DED of the four magnetic coupling structures in the figure are 180 mm. In addition, in order to more significantly display the changes in the coupling coefficient of the magnetic coupling structure under long air gaps and low coupling coefficients, the ordinate adopts logarithmic form.

It can be seen from Fig. 4.29 that under a long air gap greater than 90 mm, the CSC can significantly enhance the coupling degree of the magnetic coupling structure. Among them, when D = 90 mm, the coupling coefficient of the planar solenoid is similar to the proposed magnetic structure, but as the distance increases, its coupling coefficient decreases sharply. Compared with the hollow three-dimensional solenoid, the coupling coefficient of the improved structure is always at a higher level of 2–3 times. Therefore, the CSC effectively improves the coupling degree between coils and is suitable for long air gap WPT systems.

In addition, Fig. 4.29 compares the CSC with the three-dimensional magnetic cubic coupling structure proposed in Sects. 4.1 and 4.2 of this chapter. It can be clearly seen that the WPT distance-coupling coefficient curves of the CSC proposed

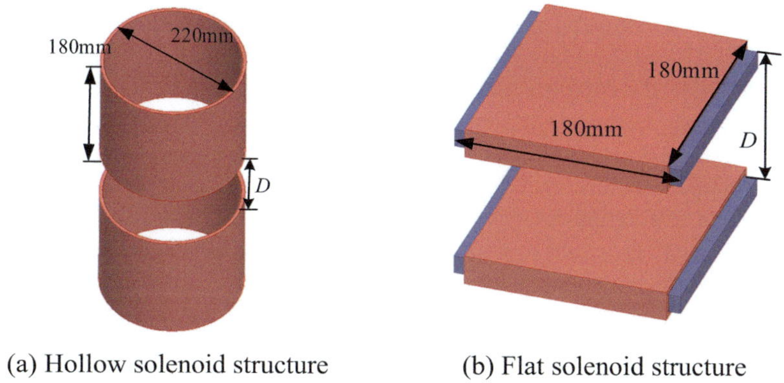

Fig. 4.28 Two types of loosely coupled transformers before improvement

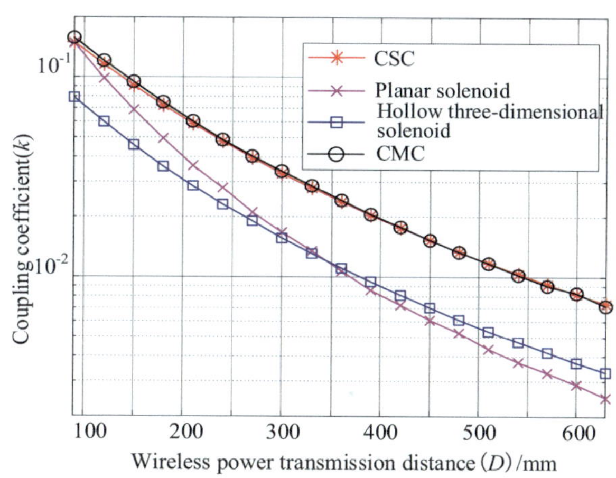

Fig. 4.29 Comparison of four loosely coupled transformers under long air gap WPT

in this section and the CMC under the same size tend to be consistent. This shows that under long air gap conditions, the long air gap WPT characteristics of CMC and the CSC of the same size (equivalent diameter) are basically consistent. Different three-dimensional shapes have slightly different representations in different application scenarios, and their cost and other factors are also different, but their essential electrical characteristics tend to be the same.

4.4 Conclusion

This chapter proposes a 3D magnetic cubic coupling structure (Cubic Magnetic Coupler, CMC) and an improved 3D solenoid magnetic coupling structure (Cylindrical Solenoid Coupler, CSC) applied to long air gap WPT systems.

When applied to long air gap WPT systems, the CMC can still maintain sufficient coupling coefficients and has strong anti-offset capabilities. First, by describing the concept of DDR, the comparison between magnetic coupling structures of different sizes and shapes is made simpler and fairer. Next, CMC was proposed, and the method of longitudinal splitting of the magnetic core was used to optimize the shape of the magnetic structure, thereby reducing its weight and cost. Multi-dimensional finite element evaluation analysis is used to optimize the size parameters of the magnetic coupling structure to improve its coupling level. Then, the proposed CMC is compared with six existing magnetic coupling structures, and evaluated from two dimensions: long air gap coupling coefficient characteristics and anti-offset capability. Finally, the bilateral LCC compensation topology was selected to perform reactive power compensation on the coil, and a CMC prototype with an outer size of $180 \times 180 \times 180\ mm^3$ was produced, and based on this, a WPT system prototype was constructed. The superior performance of the proposed CMC was verified through a series of experiments. This magnetic structure can effectively increase the coupling coefficient between the primary and secondary coils under the same DDR, and has good anti-offset capability.

Based on the analysis, design and optimization methods of the CMC, an improved CSC is proposed. The shape and parameters of the magnetic coupling structure have been optimized. The longitudinal subdivision method of the magnetic core can make it more conducive to practical applications and solve the need for core curvature for magnetic structures in cylindrical shapes. Compared with the improved planar solenoid and hollow 3D solenoid, the CSC can still maintain a sufficient coupling coefficient under long air gap conditions to meet the needs of WPT.

References

1. Gu P, Wang Y, Mai J, Yao Y, Gao S, Huang S, Xu D (2022) A three-stage-five-coil IPT system based on cylindrical solenoid coupler applied to state detection equipment of HV device. IEEE Trans Power Electron 37(2):2382–2393
2. Deng Q, Liu J, Czarkowski D et al (2016) Frequency-dependent resistance of litz-wire square solenoid coils and quality factor optimization for wireless power transfer. IEEE Trans Ind Electron 63(5):2825–2837
3. Zhou H, Gao X, Lai J et al (2018) Natural frequency optimization of wireless power systems on power transmission lines. IEEE Access 6:14038–14047
4. Choi BG, Kim JH, Lee ES et al (2019) Optimal dipole-coil ampere-turns design for maximum power efficiency of IPT. IEEE Trans Power Electron 35(7):7317–7327
5. Covic GA, Boys JT (2013) Modern trends in inductive power transfer for transportation applications. IEEE J Emerg Sel Top Power Electron 1(1):28–41

Chapter 5
Long-Range Wireless Power Transfer System Based on Double-Layer Compensation Methods

Abstract A 2.5 m five-stage IPT system based on domino cylindrical solenoid coupler compensated respectively in layers is proposed in this Chapter. The distributed series compensation method is applied to double-layer cylindrical solenoid coupler. Compared with the two types of parallel schemes, this method can significantly reduce the current of the main circuit. Compared with the concentrated series compensation method, this method can effectively reduce the resistance caused by the stray capacitance. The circuit model of the system with constant voltage output characteristics is analyzed. A prototype with a maximum efficiency of 66.7% is built. When the power is transferred to a larger value of 77.8 W, the efficiency can still reach 62%. Comparing with previous works, the wireless distance is extended while the efficiency is improved.

Keyword Long-range Inductive power transfer · Magnetic coupler design · Stray capacitance

5.1 Double-Layer Solenoidal Coupling Structure Design

In the research in Chap. 4, a 3D Cubic Magnetic Coupler (CMC) and an improved 3D Cylindrical Solenoid Coupler (CSC) are proposed for application in long air gap WPT systems. The CMC and CSC can still maintain sufficient coupling coefficient to meet WPT requirements under long air gap conditions. In order to further extend the distance of WPT, it is necessary to improve the efficiency of the WPT system by optimizing the magnetic coupling structure of the three-dimensional solenoid.

As shown in Fig. 5.1, it is an optimized single double-layer three-dimensional solenoid. The Litz coil is fixed on the acrylic shell, and the inner side of the inner coil is covered with a ferrite core designed by the longitudinal splitting method of the core. The single coil consists of an inner coil, an outer coil, an acrylic structure and a ferrite core.

Among them, the dimensions of each part in Fig. 5.1a are shown in Table 5.1.

Y. Wang et al., *Long Range Inductive Power Transfer*, CPSS Power Electronics Series,
https://doi.org/10.1007/978-981-96-7541-8_5

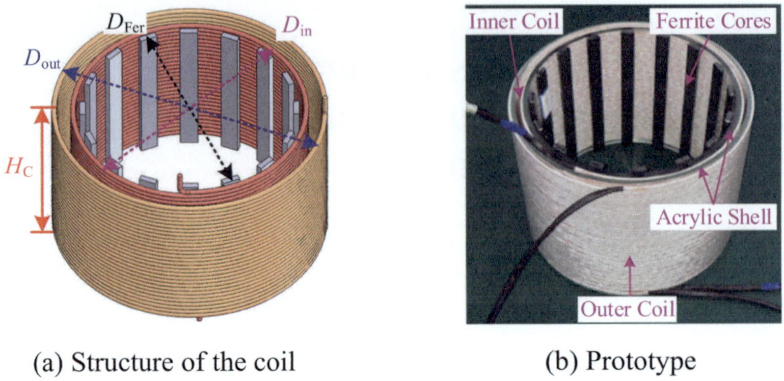

Fig. 5.1 Structure and parameters of double-layer cylindrical solenoid coupler

Table 5.1 Parameters of double-layer cylindrical solenoid coupler

Parameter	Physical meaning	Value	Parameter	Physical meaning	Value
D_{out}	Outer coil diameter	240–330 mm	n_{out}	Number of outer coil turns	28
D_{in}	Inner coil diameter	230 mm	n_{in}	Number of inner coil turns	28
H_C	Height of inner and outer coils	180 mm		Ferrite core parameters	180 * 15 * 5 mm^3
D_{Fer}	Core surrounding diameter	100–210 mm	n_{Fer}	Number of ferrite core groups	16

In a single double-layer CSC with the same turn number of the inner and outer coils, the outer coil has a larger diameter, which helps increase the inductance. At the same time, the inner coils are closer to the ferrite cores, which also tends to increase inductance. In order to carry out the following works, D_{out} and D_{Fer} need to be designed to make the self-inductance values of the inner and outer coils approximately equal through the finite element analysis (FEA) program ANSYS MAXWELL as shown in Fig. 5.2.

In Fig. 5.2, δ_L is the self-inductance value difference rate of the inner and outer coils, which is defined as follows:

$$\delta_L = 100\% \times \frac{L_O - L_I}{L_I} \tag{5.1}$$

where L_I and L_O are the self-inductances of the inner and outer coils of a single CSC. In the green area of the figure, δ_L is within 2.5%, which means that the self-inductances of the inner and outer coils are approximately equal. In addition, the

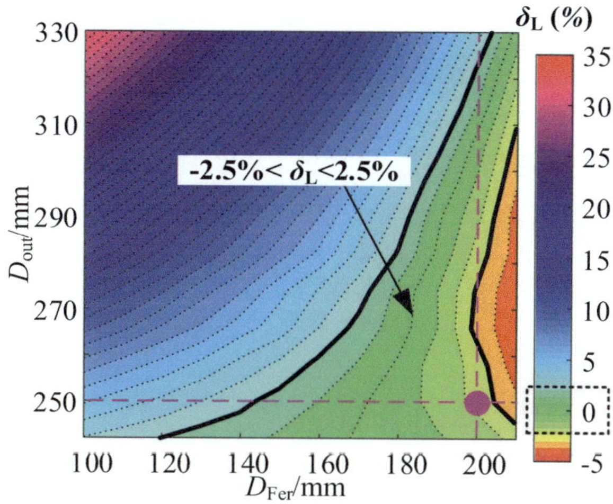

Fig. 5.2 Design of D_{out} and D_{Fer}

diameter of the outer coil should not be too large due to size constraints, while the ferrite core should be close to the inner coil for better coupling. Therefore, D_{out} and D_{Fer} are designed as 250 mm and 200 mm as shown by the magenta dots in the figure with δ_L is -1.6%.

From the magnetomotive force Eq. (5.2), it can be seen that the number of turns N of double-layer CSC has doubled, when the inner coil and outer coil pass the same current, the magnetomotive force F will be greatly increased, which helps to improve the capability inductive power transfer. In addition, the nested double-layer CSC will not take up more space, which is conducive to practical applications.

$$F = NI \tag{5.2}$$

FEA model for two identical double-layer CSCs has been built in Fig. 5.3a, k_{IO} is the coupling coefficient of the inner coil and the outer coil in the same coupler. $k_{I\text{-}I}$, $k_{O\text{-}O}$, $k_{I\text{-}O}$ are the coupling coefficients between the inner-inner, outer-outer and inner-outer coils in different couplers. The variation of coupling coefficients with the distance-to-diameter ratio (DDR) of the two couplers is shown in Fig. 5.3b. Among them, DDR in (5.3) reflects the relative distance of IPT between the two couplers, D_{couple} is the power transfer distance between the two couplers and d_{ED} is the equivalent diameter of single coupler.

$$\text{DDR} = \frac{D_{couple}}{d_{ED}} \tag{5.3}$$

It can be seen from Fig. 5.3b that k_{IO} is always approximately equal to 1, which shows that the two coils in the same coupler are in an extremely tightly coupled state. Besides, the curves of $k_{I\text{-}I}$, $k_{O\text{-}O}$, $k_{I\text{-}O}$ are basically coincident, and the coils between the two couplers can still maintain an acceptable coupling coefficient greater than 0.01 when the relative distance of wireless power transfer is large enough.

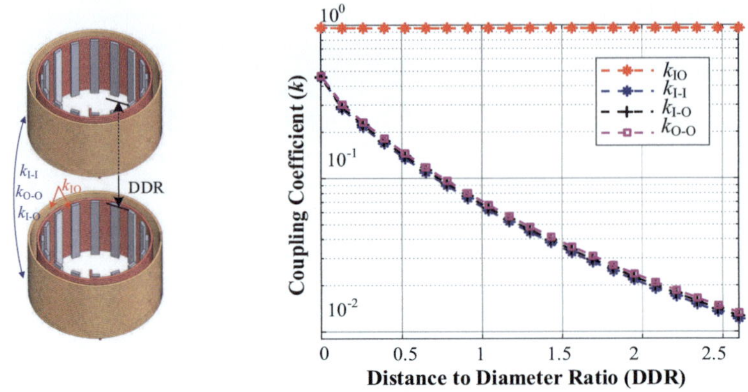

(a) Simulation Model (b) Trend of coupling coefficient with DDR variation

Fig. 5.3 Finite element analysis

5.2 Study of Connection and Compensation Schemes for Magnetically Coupled Structures

From the above analysis, it can be found that the double-layer three-dimensional solenoid magnetic coupling structure can multiply the magnetomotive force generated by the magnetic structure when the same current flows through the wire. Parallel connection and series connection are the two basic connection methods of double-layer coils. This section will analyze, design and compare the characteristics of double-layer magnetic coupling structures under different connection and compensation methods, and select the most efficient method as a solution to achieve long air gap wireless power transmission.

5.2.1 Parallel Compensation Methods

For the primary and secondary side double-layer CSCs with parallel method, an ideal model as shown in Fig. 5.4a is established. L_{PI}, L_{PO}, L_{SI} and L_{SO} are the self-inductance of each coil, M_{PISI}, M_{PISO}, M_{POSI}, M_{PISI} are the mutual inductance of each coil between different CSCs, and M_{PIO}, M_{SIO} are the mutual inductance between the inner coil and the outer coil in the same CSC. According to Fig. 5.3b, when the value of DDR is large enough, the coupling coefficient between the two CSC coils drops below 0.1, while k_{IO} is always close to 1. Therefore, the influence of mutual inductance between CSCs can be ignored while analyzing a single double-layer CSC. Therefore, according to the characteristics of the parallel coupled inductor, Fig. 5.4a could be equivalent to Fig. 5.4b. The self-inductance values are calculated by the following equations:

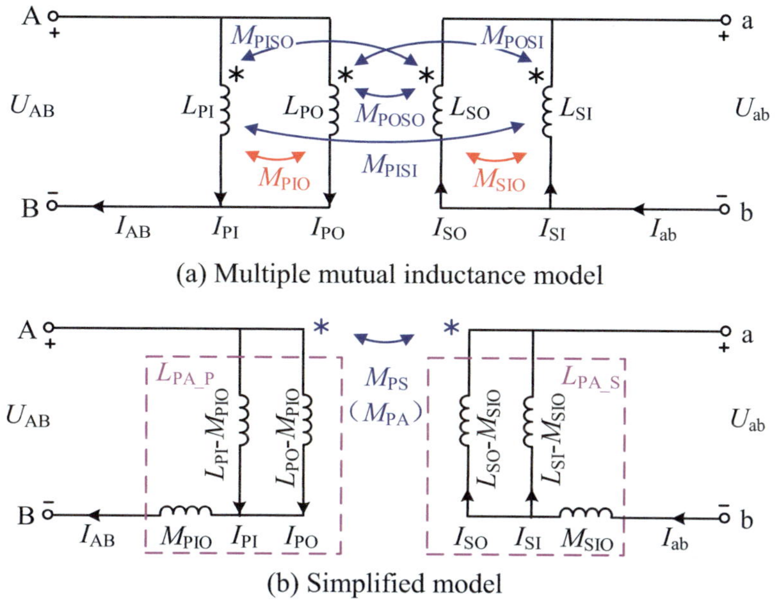

Fig. 5.4 Parallel circuit model

$$L_{PA\_P} = \frac{L_{PI}L_{PO} - M_{PIO}^2}{L_{PI} + L_{PO} - 2M_{PIO}} \tag{5.4}$$

$$L_{PA\_S} = \frac{L_{SI}L_{SO} - M_{SIO}^2}{L_{SI} + L_{SO} - 2M_{SIO}} \tag{5.5}$$

The concentrated parallel compensation method (*Type I*) to achieve zero phase angle (ZPA) as shown in Fig. 5.5, $L_{I\text{-}P} = L_{PA\text{-}P}$, $L_{I\text{-}S} = L_{PA\text{-}S}$. $C_{I\text{-}P}$, $C_{I\text{-}S}$ are the compensation capacitors on both sides. It is calculated (5.6) and (5.7).

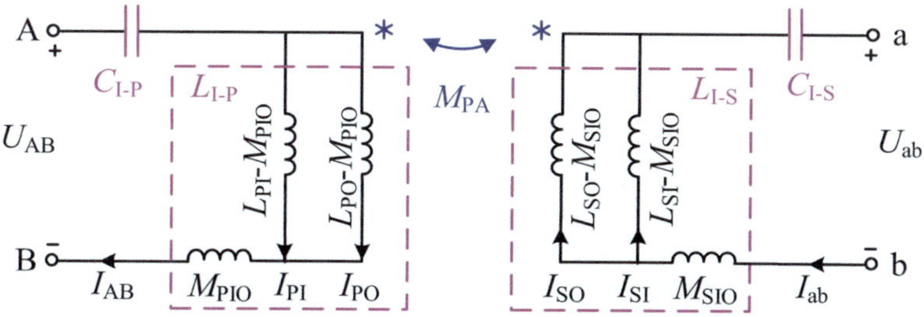

Fig. 5.5 Circuit model of the concentrated parallel compensation method

$$C_{\text{I-P}} = 1/\omega^2 L_{\text{I-P}} \tag{5.6}$$

$$C_{\text{I-S}} = 1/\omega^2 L_{\text{I-S}} \tag{5.7}$$

In Fig. 5.2, the self-inductance of the inner and outer layers is basically the same in the green area. However, there will be a tiny difference in self-inductance between the inner and outer coils due to practical factors. Taking the parallel double-layer CSC on the primary side as an example, the current imbalance ratio λ of the currents flowing through the two coils can be defined by (5.8)–(5.11) according to Fig. 5.5.

$$\lambda = \frac{I_{\text{PI}}}{I_{\text{PO}}} = \frac{L_{\text{PO}} - M_{\text{PIO}}}{L_{\text{PI}} - M_{\text{PIO}}} \tag{5.8}$$

$$k_{\text{PIO}} = \frac{M_{\text{PIO}}}{\sqrt{L_{\text{PI}} L_{\text{PO}}}} \tag{5.9}$$

$$\alpha_{\text{PIO}} = \sqrt{\frac{L_{\text{PI}}}{L_{\text{PO}}}} \tag{5.10}$$

$$\lambda = \frac{1 - k_{\text{PIO}}\sqrt{\frac{L_{\text{PI}}}{L_{\text{PO}}}}}{\frac{L_{\text{PI}}}{L_{\text{PO}}} - k_{\text{PIO}}\sqrt{\frac{L_{\text{PI}}}{L_{\text{PO}}}}} = \frac{1 - k_{\text{PIO}}\alpha_{\text{PIO}}}{\alpha_{\text{PIO}}^2 - k_{\text{PIO}}\alpha_{\text{PIO}}} \tag{5.11}$$

Through numerical analysis methods and setting reasonable calculation steps, the above λ could be represented by Fig. 5.6. Where k_{PIO} is close to 1, indicating that M_{PIO} is similar to L_{PI} and L_{PO}. Ideally, the self-inductance of the coils is equal as $L_{\text{PI}} \approx L_{\text{PO}}$, and $I_{\text{PI}} \approx I_{\text{PO}}$ can be obtained. However, the tiny difference α_{PIO} in self-inductance of the coils between L_{PI} and L_{PO} will cause a great imbalance between I_{PI} and I_{PO}, which will further increase the loss of the actual coupler. In the actual circuit, due to ESR of the components, the larger λ values in Fig. 5.6 will be reduced by the effect of resistance.

In order to reduce the adverse effects of current imbalance, a distributed parallel compensation method (*Type II*) is applied for the double-layer CSCs as shown in Fig. 5.7a.

In Fig. 5.7a, since the self-inductance values of the inner and outer coils are nearly the same analyzed, $L_{\text{PI}} \approx L_{\text{PO}} = L_{\text{II-P}}$, $L_{\text{SI}} \approx L_{\text{SO}} = L_{\text{II-S}}$ could be assumed. In order to balance the currents, $Z_{\text{II-PI}} = Z_{\text{II-PO}}$ and $Z_{\text{II-SI}} = Z_{\text{II-SO}}$. Therefore, $C_{\text{II-PI}} \approx C_{\text{II-PO}} = C_{\text{II-P}}$, $C_{\text{II-SI}} \approx C_{\text{II-SO}} = C_{\text{II-S}}$. Besides, impedance angle should be constrained as$\angle Z_{\text{in-P}} = \angle Z_{\text{in-S}} = 0°$ to realize ZPA. Therefore, the values of compensation capacitors obtained by (5.12), (5.13), and the corresponding circuit is shown in Fig. 5.7b.

$$C_{\text{II-PI}} \approx C_{\text{II-PO}} = C_{\text{II-P}} = \frac{1}{\omega^2 (L_{\text{II-P}} + M_{\text{PIO}})} \tag{5.12}$$

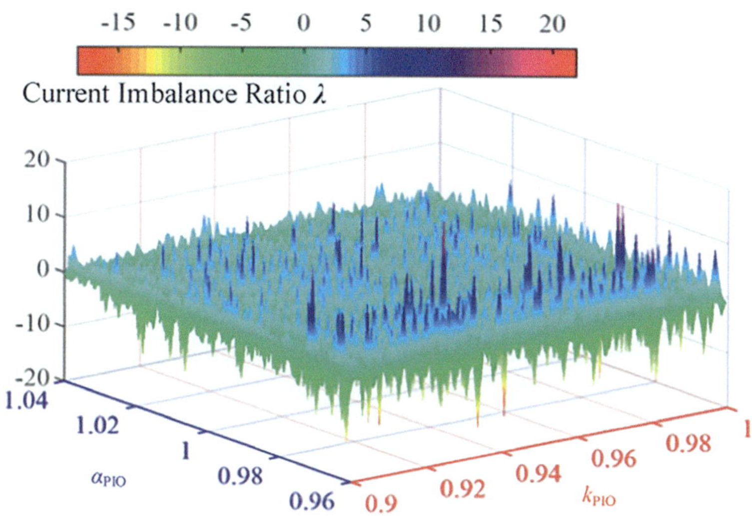

Fig. 5.6 The current imbalance ratio between the double-layer coils when the self-inductance values are not exactly the same

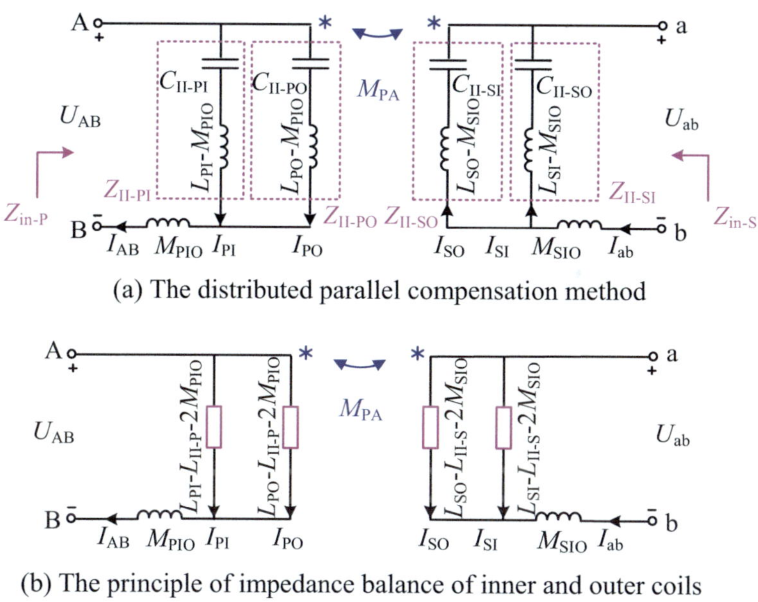

Fig. 5.7 Circuit model of the coils in distributed parallel compensation method

$$C_{\text{II-SI}} \approx C_{\text{II-SO}} = C_{\text{II-S}} = \frac{1}{\omega^2(L_{\text{II-S}}+M_{\text{SIO}})} \tag{5.13}$$

It can be found from the above analysis that (L_{PI}-$L_{\text{II-P}}$, L_{PO}-$L_{\text{II-P}}$) and (L_{SI}-$L_{\text{II-S}}$, L_{SO}-$L_{\text{II-S}}$) are very small values compared to $-2M_{\text{PIO}}$ and $-2M_{\text{SIO}}$ among the four branches impedance shown in Fig. 5.7b. Therefore, the current values of the parallel branches can be considered the same, $I_{\text{PI}} \approx I_{\text{PO}}$, $I_{\text{SI}} \approx I_{\text{SO}}$. Finally, the problem of current imbalance in the coils of double-layer parallel connection type could be solved by distributed compensation method.

In addition [1, 2], introduced the influence of magnetic components such as ferrite on the equivalent resistance of the Litz coil, which determines that the equivalent resistance of the parallel double-layer CSC cannot be reduced intensely as ideal components. In summary, compared with the single-layer CSC, the parallel double-layer coil will not greatly reduce the copper loss.

5.2.2 Series Compensation Methods

The case of double-layer CSCs in series is shown in Fig. 5.8a, the simplified model shown in Fig. 5.8b is obtained according to the above method in Fig. 5.4. When the self-inductance of the inner and outer coils is roughly the same and k_{IO} is close to 1 as shown in Fig. 5.3, the equivalent self-inductance $L_{\text{SE\_P}}$ and $L_{\text{SE\_S}}$ of the double-layer CSC can be increased to about 4 times of the single-layer case. Obviously, according to (5.2), the increase in self-inductance is conducive to reducing the current and further reducing the loss of the system when the same power is transferred.

The concentrated series compensation method (*Type III*) is shown in Fig. 5.9. The compensation capacitance value is calculated from (5.14) to (5.15). Where $L_{\text{III-P}}$ and $L_{\text{III-S}}$ are the measured total inductance values of the double-layer coils connected in series.

$$C_{\text{III-P}} = 1/\omega^2 L_{\text{III-P}} \tag{5.14}$$

$$C_{\text{III-S}} = 1/\omega^2 L_{\text{III-S}} \tag{5.15}$$

Figure 5.9b is a schematic diagram of the primary coil resonant circuit, which is a cross-sectional view of the double-layer three-dimensional solenoid shown in Fig. 5.1a. When coils are connected in series, the inductance increases and the equivalent series resistance of the magnetic coupling structure increases accordingly. In addition, the centralized series compensation method will generate additional losses between the two layers of coils caused by the higher voltage difference between the inner and outer coils and stray capacitance and parasitic resistance.

In Fig. 5.9, the equivalent AC voltage U_{AB} is added between point A and point B in the two-stage system circuit. Taking the primary side stage as an example, in

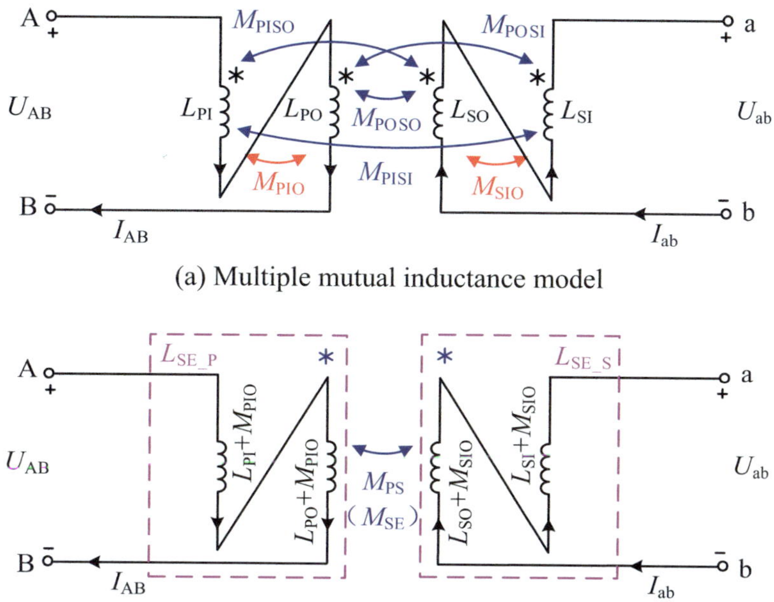

(a) Multiple mutual inductance model

(b) Simplified model

Fig. 5.8 Series circuit model

Fig. 5.9b, the current first flows through the inner coil, and then flows through the outer coil through the series connection of the coils. Since both the inner and outer coils are inductive, a higher voltage U_{gap_III} will be generated between the inner and outer coils. The following is an analysis of U_{gap_III}:

Taking point B as the zero potential reference point, when voltage U_{AB} is applied between point A and B, in Fig. 5.9c, taking the primary side as an example, when current I_{AB} flows, the compensation capacitor $C_{III\text{-}P}$ The potential drop is U_{CIII}, which can be obtained from Formula (5.16):

$$U_{CIII} = I_{AB}\frac{1}{j\omega C_{III\text{-}P}} \tag{5.16}$$

At this time, the potential of point M is $U_{M\_III} = U_{AB}\text{-}U_{CIII}$. Since the parameters of the inner and outer coils are approximately the same, the potential drops on the two coils are also approximately the same, so the potential at point N is $U_{N\_III} \approx (U_{AB}\text{-}U_{CIII})/2$.

Combined with the above potential relationship, the potential difference U_{gap_III} between points M and N can be obtained by Eq. (5.17):

$$U_{gap\_III} = U_{M\_III} - U_{N\_III} = \frac{(U_{AB} - U_{CIII})}{2} \tag{5.17}$$

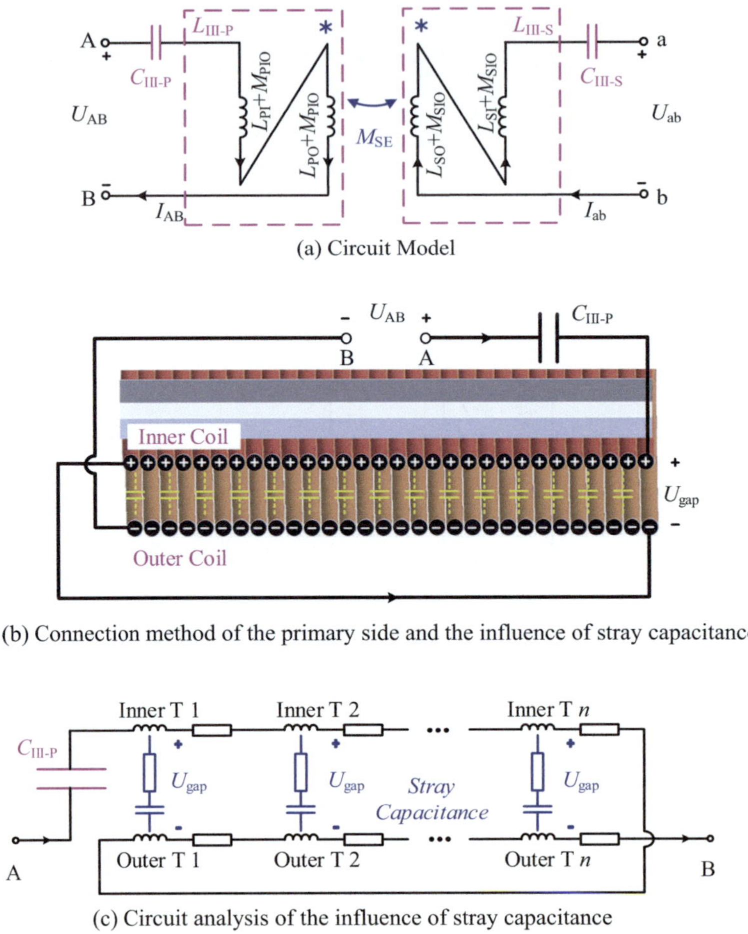

Fig. 5.9 Concentrated series compensation method

When two coils have the same turn numbers n and self-inductance value, the voltage between each turn of the inner coil and the corresponding outer coil is also U_{gap_III} shown in Fig. 5.9c. As the shape of the coils are symmetrical and nested cylinders, which can be regarded as positive and negative capacitor plates, and there is always the influence of stray capacitance between the two coils. Under the excitation of U_{gap_III}, stray currents flow through the capacitance, and additional losses are generated.

In order to eliminate the extra loss caused by stray capacitance, the distributed series compensation method (*Type IV*) shown in Fig. 5.10 is used in the IPT system based on double-layer CSC. The following takes the primary side to illustrate.

Next, in Fig. 5.10b, c, the primary side stage is taken as an example for analysis. Among them, the capacitors $C_{IV\text{-}PI}$ and $C_{IV\text{-}PO}$ are connected in series before the

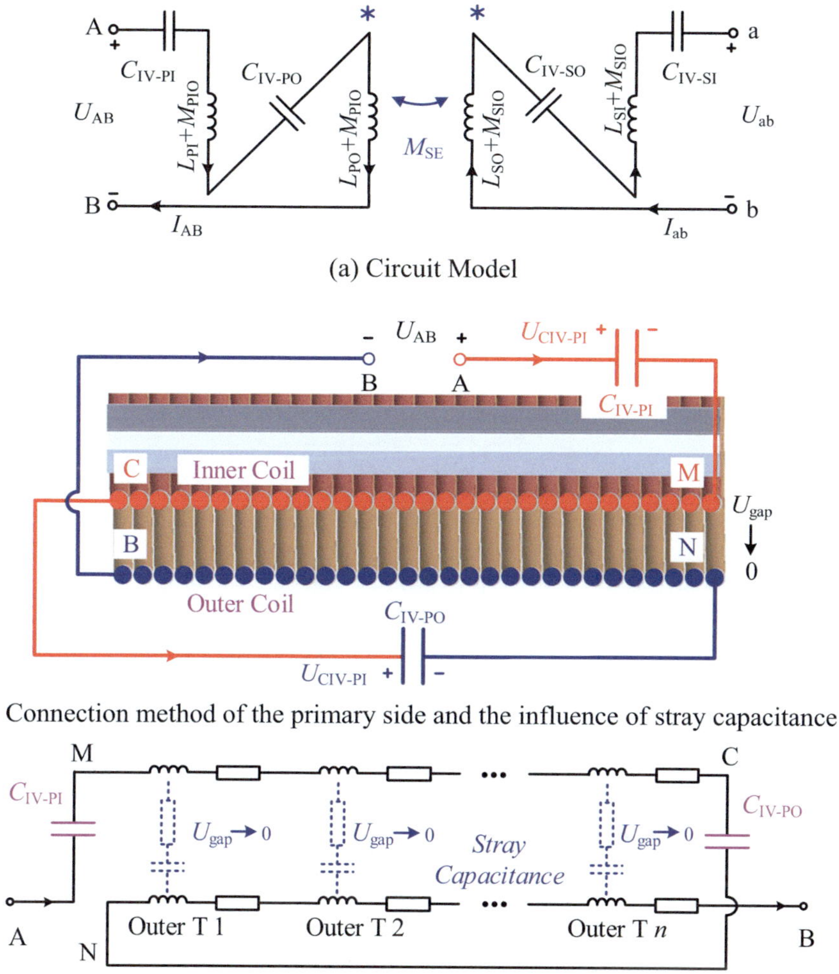

(c) Circuit analysis of the influence of stray capacitance

Fig. 5.10 Distributed series compensation method

inner and outer coils respectively. L_{PI} and L_{PO} are the self-inductance of the inner and outer coils respectively, and M_{PIO} is the mutual inductance between the two coils. The calculation formulas for capacitance parameters are as follows:

Among them, $C_{IV\text{-}PI}$ is the series compensation capacitor of the inner coil, which resonates in series with $L_{PI} + M_{PIO}$, and $C_{IV\text{-}PO}$ is in series resonance with $L_{PI} + M_{PIO}$.

In Fig. 5.10b, after being compensated by distributed series, the red circuit A-M-C and the blue circuit C-N-B are in resonance respectively. Besides, since $L_{PI} \approx L_{PO}$ analyzed in Sect. 5.1, $L_{PI} + M_{PIO} \approx L_{PO} + M_{PIO}$, the values of capacitors $C_{IV\text{-}PI}$ and $C_{IV\text{-}PO}$ in (5.18) and (5.19) are substantially equal. Therefore, in a series circuit, since the current flowing through the compensation capacitor is the same, it can be concluded that the voltage across the compensation capacitor $U_{CIV} = U_{CIV\text{-}PI} \approx U_{CIV\text{-}PO}$. At this time, the circuit model of the double-layer coil stray loss is shown in Fig. 5.10c. According to the afore mentioned centralized series compensation analysis method, point B is used as the reference point, the potential of point A is U_{AB}, and the potential of point M is $U_{M\_IV} = U_{AB}\text{-}U_{CIV}$. Due to the symmetry of the circuit model, the potential of point C in the circuit is $U_{C\_IV} = U_{AB}/2$. Similarly, the potential of point N is $U_{N\_IV} = U_{C\_IV} - U_{CIV}$, and the potential difference between points M and N is $U_{gap\_IV} = U_{M\_IV} - U_{N\_IV} = U_{AB}/2$.

$$C_{IV\text{-}PI} = \frac{1}{\omega^2 L_{IV\text{-}PI}} = \frac{1}{\omega^2 (L_{PI} + M_{PIO})} \tag{5.18}$$

$$C_{IV\text{-}PO} = \frac{1}{\omega^2 L_{IV\text{-}PO}} = \frac{1}{\omega^2 (L_{PO} + M_{PIO})} \tag{5.19}$$

Generally, the voltage stress U_{CIII} of the series compensation capacitor in a weakly coupled WPT system is much greater than the input voltage U_{AB}. Therefore, it can be seen from the following Eq. (5.20) that compared with U_{gap_III} obtained by using the centralized series compensation method, the interlayer voltage U_{gap_IV} between the double-layer coils based on the distributed series compensation method is significantly reduced. Therefore, the loss caused by inter-layer stray capacitance is also greatly reduced, and the efficiency of the WPT system will be improved.

$$U_{gap\_III} = \left| \frac{(U_{AB} - U_{CIII})}{2} \right| \gg \left| \frac{U_{AB}}{2} \right| = U_{gap\_IV} \tag{5.20}$$

In addition, since the number of turns of the inner and outer coils is the same, it can also be ensured that the potential difference between the corresponding turns of the inner and outer coils under the distributed series compensation method is much smaller than that under the centralized series compensation method. This shows that almost no current flows through the stray capacitance in Fig. 5.10c, which can greatly reduce the additional loss caused by stray capacitance.

5.2.3 Experimental Analysis and Comparison

In order to select the best connection and compensation scheme to effectively improve the efficiency of the IPT system based on double-layer CSC, four sets of circuits with different connection and compensation methods have been completed, named *Type I*, *Type II*, *Type III* and *Type IV* respectively.

Type I: Concentrated parallel compensation method for double-layer CSCs shown in Fig. 5.5.

Type II: Distributed parallel compensation method for double-layer CSCs shown in Fig. 5.7.
Type III: Concentrated series compensation method shown in Fig. 5.9.
Type IV: Distributed series compensation method shown in Fig. 5.10.

The circuit model of the wireless power transmission system based on the double-layer three-dimensional solenoid magnetic coupling structure under the four-group connection and compensation scheme is shown in Fig. 5.11. The system consists of a DC power supply, an inverter circuit module, a wireless power transmission module including a magnetic structure and four sets of connection compensation schemes, a rectifier circuit module and a load R.

Based on two double-layer CSCs shown in Fig. 5.1, a two-stage IPT system was built in Fig. 5.12, and reactive power compensation is performed under the conditions of the above four types. During the experiment, only the connection compensation scheme of the double-layer magnetic structure of this system was replaced from *Type I* to *Type IV*.

The wireless distance of the high DDR systems is 0.4 m, and DDR = 2.04. Parameters in Fig. 5.11 are measured by impedance analyzer KEYSIGHT E4990A shown in Table 5.2. Among them, $L_{\text{I-P}}$, $L_{\text{I-S}}$, $L_{\text{II-PI}}$, $L_{\text{II-PO}}$, $L_{\text{II-SI}}$, $L_{\text{II-SO}}$ $L_{\text{III-P}}$, $L_{\text{III-S}}$, $L_{\text{IV-PI}}$, $L_{\text{IV-PO}}$, $L_{\text{IV-SI}}$ and $L_{\text{IV-SO}}$ are the self-inductances of the corresponding coils referred to in Figs. 5.5, 5.7, 5.9, and 5.10. $C_{\text{I-P}}$, $C_{\text{I-S}}$, $C_{\text{II-PI}}$, $C_{\text{II-PO}}$, $C_{\text{II-SI}}$, $C_{\text{II-SO}}$ $C_{\text{III-P}}$, $C_{\text{III-S}}$, $C_{\text{IV-PI}}$, $C_{\text{IV-PO}}$, $C_{\text{IV-SI}}$ and $C_{\text{IV-SO}}$ are the series resonant capacitors calculated by (5.6), (5.7), (5.12), (5.13), (5.14), (5.15), (5.18), (5.19). M_{PIO}, M_{SIO} are the mutual inductance between the inner coil and the outer coil of the primary and secondary

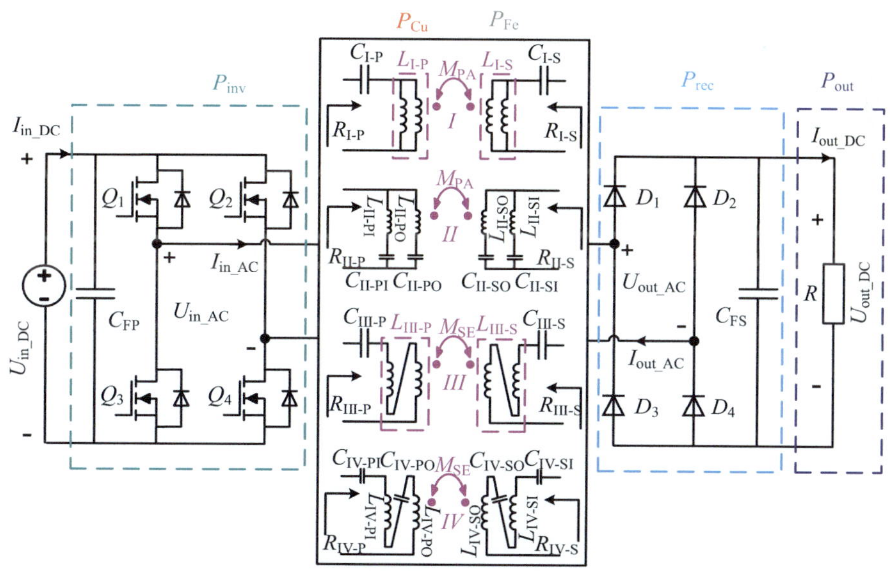

Fig. 5.11 Circuit model of the IPT system under the four connection and compensation scheme

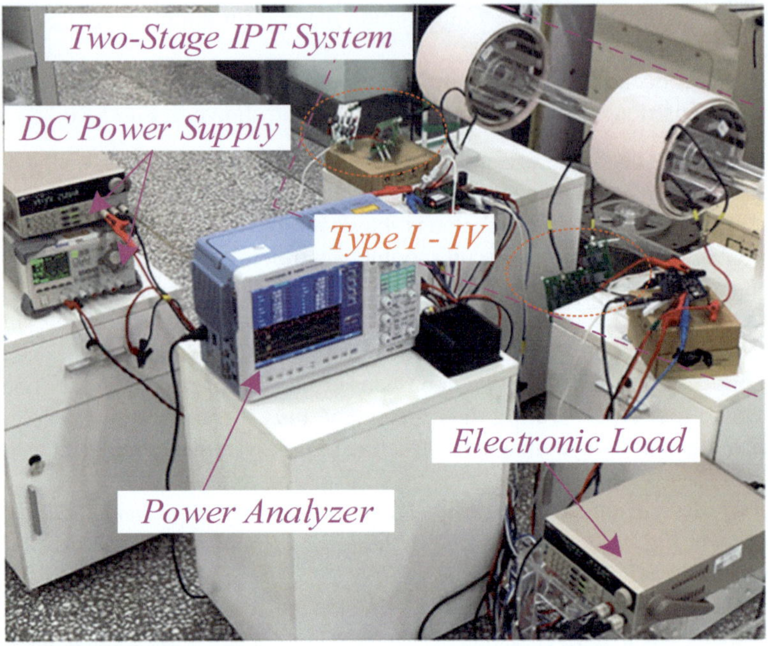

Fig. 5.12 The two-stage prototype

CSC. M_{PA} and M_{SE} are the mutual inductance between the primary coil and the secondary coil in parallel and series. $R_{I\text{-}P}$, $R_{II\text{-}P}$, $R_{III\text{-}P}$ and $R_{IV\text{-}P}$ are the AC equivalent series resistance of the primary side in the case of the four types, and $R_{I\text{-}S}$, $R_{II\text{-}S}$, $R_{III\text{-}S}$ and $R_{IV\text{-}S}$ are the secondary side cases. $R_{IV\text{-}P}$ and $R_{IV\text{-}S}$ are significantly smaller than $R_{III\text{-}P}$ and $R_{III\text{-}S}$, which shows that the distributed series compensation method effectively reduces the equivalent AC resistance of the IPT system.

The voltage, current, power and efficiency of the system are analyzed with power analyzer YOKOGAWA PX8000 under the four types shown in Table 5.3.

In the four sets of control experiments, the output power P_{out} is 25.4 W. The corresponding parts of the power loss in Table 5.3 have been marked in Fig. 5.11. It can be seen from the waveforms that the phases of the AC input voltage and current are same, which shows that the four types of circuits are all in resonance.

Among them, P_{inv} and P_{rec} are the power loss of the inverter and rectifier, which mainly depends on the currents when the types of Q_1–Q_4 and D_1–D_4 are determined. Since the I_{in_DC} and I_{out_DC} of *Type I* and *Type II* are about twice that of *Type III* and *Type IV*, the losses of *Type I* and *Type II* circuit in the inverter and rectifier part are significantly larger.

Table 5.2 Parameters of the IPT system under the four schemes

Parameter	Value/type	Parameter	Value	Parameter	Value
f	100 kHz	M_{PIO}	364.3 μH	$R_{III\text{-}S}$	1520 mΩ
Q_1–Q_4	IRFBA22N50	M_{SIO}	361.7 μH	$L_{IV\text{-}PI}$	737.3 μH
D_1–D_4	MBR40200WT	$C_{II\text{-}PI}$	3.44 nF	$L_{IV\text{-}PO}$	729.4 μH
$L_{I\text{-}P}$	370.5 μH	$C_{II\text{-}PO}$	3.46 nF	$L_{IV\text{-}SI}$	730.9 μH
$L_{I\text{-}S}$	365.7 μH	$C_{II\text{-}SI}$	3.48 nF	$L_{IV\text{-}SO}$	725.2 μH
$C_{I\text{-}P}$	6.84 nF	$C_{II\text{-}SO}$	3.49 nF	$C_{IV\text{-}PI}$	3.44 nF
$C_{I\text{-}S}$	6.93 nF	$R_{II\text{-}P}$	219 mΩ	$C_{IV\text{-}PO}$	3.47 nF
$R_{I\text{-}P}$	280 mΩ	$R_{II\text{-}S}$	213 mΩ	$C_{IV\text{-}SI}$	3.47 nF
$R_{I\text{-}S}$	270 mΩ	$L_{III\text{-}P}$	1483.8 μH	$C_{IV\text{-}SO}$	3.49 nF
$L_{II\text{-}PI}$	371.3 μH	$L_{III\text{-}S}$	1460.5 μH	$R_{IV\text{-}P}$	690 mΩ
$L_{II\text{-}PO}$	366.8 μH	$C_{III\text{-}P}$	1.71 nF	$R_{IV\text{-}S}$	670 mΩ
$L_{II\text{-}SI}$	367.1 μH	$C_{III\text{-}S}$	1.73 nF	M_{PA}	7.18 μH
$L_{II\text{-}SO}$	363.6 μH	$R_{III\text{-}P}$	1540 mΩ	M_{SE}	29.14 μH

P_{Cu} is the loss caused by the series equivalent resistance of the primary and secondary sides of the AC circuit. The resistances $R_{III\text{-}P}$ and $R_{III\text{-}S}$ of *Type III* are significantly greater than $R_{IV\text{-}P}$ and $R_{IV\text{-}S}$ of *Type IV*. Therefore, in the case of similar currents, the P_{Cu} of *Type III* is significantly larger. In addition, compared with the *Type I*, the unbalance of the current flowing through the parallel coils is effectively reduced in *Type II*. Therefore, P_{Cu} of *Type II* is smaller than *Type I*, and its proportion is similar to that of *Type IV*.

The loss caused by the series equivalent resistance of the secondary AC circuit (copper loss). Among them, the resistances $R_{III\text{-}P}$ and $R_{III\text{-}S}$ of *Type III* are significantly larger than those of *Type IV* $R_{IV\text{-}P}$ and $R_{IV\text{-}S}$. Therefore, when the current magnitude is similar, the copper loss of type III is significantly larger. In addition, compared with *Type I*, *Type II* effectively reduces its P_{Cu} by balancing the current of the inner and outer coils, and its proportion of copper loss is similar to that of *Type IV*.

P_{Fe} is the magnetic loss (iron loss) produced by the ferrite core, which is obtained from the Steinmetz equation [3] in (5.21). Substituting Ohm's magnetic circuit law in (5.22) into (5.21), we can get Eq. (5.23).

$$P_{Fe} = VC_{core}f^{\alpha}B^{\beta} \tag{5.21}$$

$$F = NI = Hl = \frac{Bl}{\mu} \tag{5.22}$$

$$P_{Fe} = VC_{core}f^{\alpha}\left(\frac{\mu}{l}\right)^{\beta}(NI)^{\beta} \tag{5.23}$$

Table 5.3 Experimental results and loss analysis of the two-stage IPT system under different connection and compensation methods

Type	Power, efficiency and waveform	Loss analysis (η)	Loss analysis (η_{fix}) without P_{add}
Type I 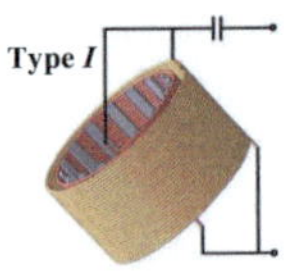	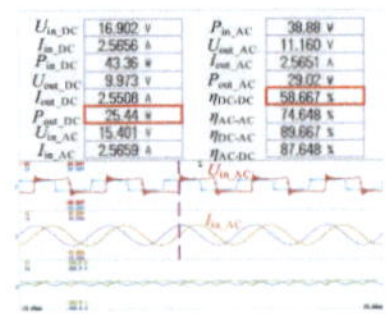	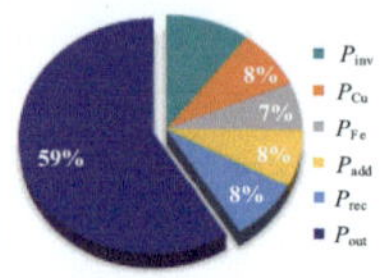	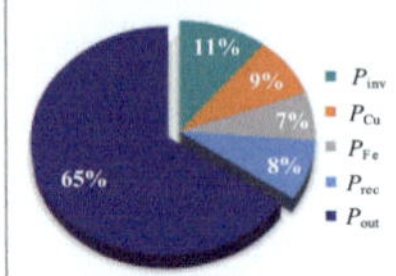
Type II 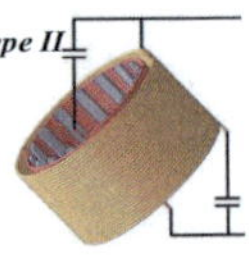	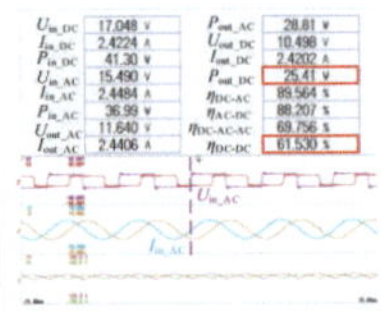	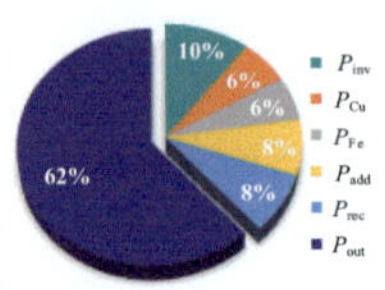	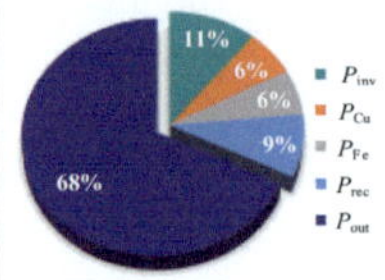
Type III 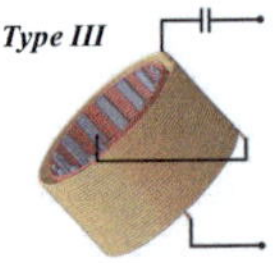	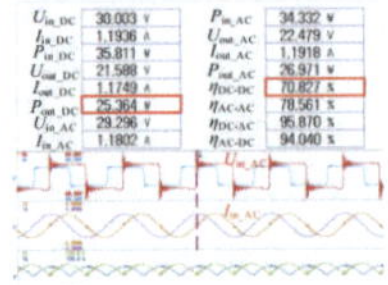	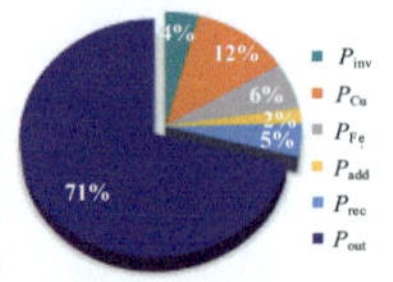	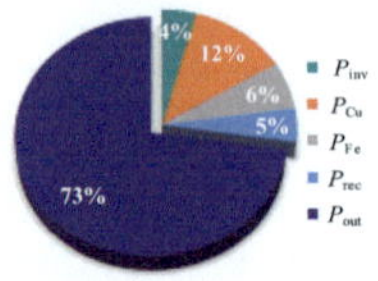
Type IV 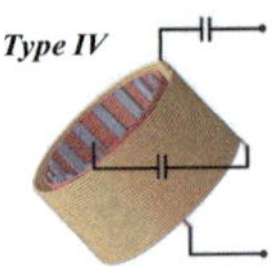	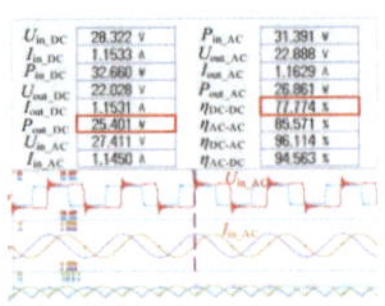	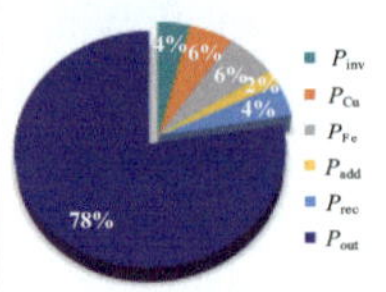	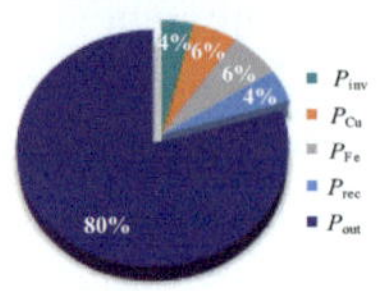

Among them, V, C_{core}, α, β, μ and l are related to the characteristics of the magnetic cores and the space where the two CSCs are located. In the four sets of experiments, these quantities are the same. Therefore, comparison the P_{Fe} here depends only on $F = NI$. Since P_{out} of the four experiments are the same, the magnetic It is worth noting that the two AC current probes of the power analyzer will be connected to the circuit in series when measuring AC current. It is equivalent to a 260 mΩ resistor connected in series on the primary and secondary AC circuits, and additional loss P_{add} is generated. Since the inverters and rectifiers of the four-type circuits are directly connected in series with the current probes, additional loss of the primary side P_{add_P} and the secondary side P_{add_S} can be equivalent to the decrease value of input power and increase value of output power respectively, and $P_{add} = P_{add\_P} + P_{add\_S}$. The efficiency calculation methods that take into account P_{add} and eliminate the influence of P_{add} are shown in (5.24) and (5.25) respectively. Among them, η_{fix} is the correction to η.

$$\eta = \frac{P_{out\_DC}}{P_{in\_DC}} = \frac{P_{out}}{P_{inv} + P_{rec} + P_{Cu} + P_{Fe} + P_{add} + P_{out}} \tag{5.24}$$

$$\eta_{fix} = \frac{P_{out} + P_{add\_S}}{P_{inv} + P_{rec} + P_{Cu} + P_{Fe} + P_{add\_S} + P_{out}} \tag{5.25}$$

In summary, currents in the main circuits of *Type I* and *Type II* are significantly larger, which will cause much more losses in the inverter, rectifier and compensation capacitors than *Type III* and *Type IV*. Comparing with *Type II*, currents of the two coils in *Type I* are unbalanced because of the slight difference between the self-inductance values of the coils. Therefore, the efficiency of *Type I* is the lowest among the four types. The distributed series compensation method has been applied in *Type IV*, which significantly reduces the ESR on AC side compared to *Type III*. Therefore, the efficiency of the proposed IPT system based on *Type IV* connection and compensation is the highest among the four cases. In the following work, *Type IV* is selected as the connection and compensation method of the couplers.

5.3 Research on Multi-stage Wireless Power Transfer System Based on Distributed Series Compensation Method

5.3.1 Circuit Modeling and Analysis of the System

The circuit model of five-stage IPT system based on domino cylindrical solenoid coupler compensated respectively in layers is shown in Fig. 5.13. The rectifier and inverter parts are the same as Fig. 5.11, (L_{AI}, L_{AO}) ~ (L_{EI}, L_{EO}) are the equivalent self-inductance of the five double-layer CSCs A ~ E, and (C_{AI}, C_{AO}) ~ (C_{EI}, C_{EO}) are the series compensation capacitors of the CSCs calculated by (5.11), M_{AB} ~ M_{DE} are the mutual inductance between every two adjacent CSCs in the five-stage system. Among them, the mutual inductance between non-adjacent double-layer CSCs has been ignored.

In order to further analyze the characteristics of the circuit, AC side circuit based on the controlled source model in Fig. 5.14 is obtained. Among them, the influence of the equivalent series resistance in the system is ignored.

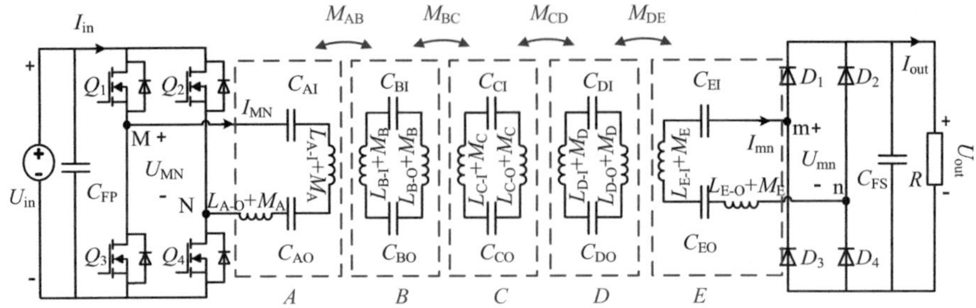

Fig. 5.13 The circuit model of five-stage IPT system

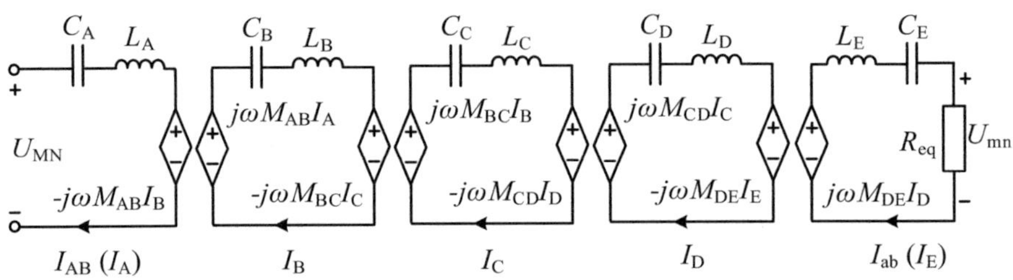

Fig. 5.14 AC equivalent circuit of the system

Zero phase angle (ZPA) of the system can be achieved when the above circuit of Fig. 5.14 meets the series resonance conditions in (5.26).

$$\frac{1}{j\omega C_i} + j\omega L_i = 0 \quad i = A, B, C, D, E \tag{5.26}$$

The reactive power in the system is the smallest and the efficiency reaches the highest in this condition. The following conclusions can be obtained:

$$\begin{cases} I_A = U_{MN}\frac{M_{BC}^2 M_{DE}^2}{R_{eq}M_{AB}^2 M_{CD}^2} \\ I_B = -U_{MN}\frac{1}{j\omega M_{AB}} \\ I_C = -U_{MN}\frac{M_{BC}M_{DE}^2}{R_{eq}M_{AB}M_{CD}^2} \\ I_D = U_{MN}\frac{M_{BC}}{j\omega M_{AB}M_{CD}} \\ I_E = U_{MN}\frac{M_{BC}M_{DE}}{R_{eq}M_{AB}M_{CD}} \end{cases} \tag{5.27}$$

Further, the constant voltage (CV) characteristics of the IPT system can be obtained.

$$G_V = \frac{U_{out}}{U_{in}} = \frac{U_{mn}}{U_{MN}} = \frac{M_{BC}M_{DE}}{M_{AB}M_{CD}} \tag{5.28}$$

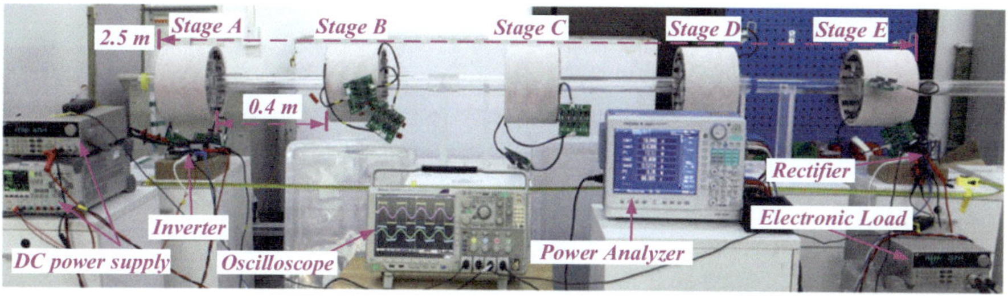

Fig. 5.15 Prototype of the system

5.3.2 Experimental Verification

Based on the domino cylindrical solenoid coupler series compensated respectively in layers, a five-stage system prototype platform with a length of 2.5 m has been built as shown in Fig. 5.15, each part of *Stage A-E* corresponds to Fig. 5.13. Among them, structure and parameters of the coils are shown in Fig. 5.1, the method of connection and compensation are shown in Fig. 5.10, and the circuit model is shown in Fig. 5.13. It is worth noting that, in order to eliminate the influence of P_{add} caused by the current probes in TABLE III, only the DC side is measured by power analyzer YOKOGAWA PX8000. For AC circuits, an oscilloscope Tektronix MDO4104C is used for measurement.

Parameters of the system in Fig. 5.13 are measured by impedance analyzer KEYSIGHT E4990A shown in Table 5.4. Among them, R_A ~ R_E are ESRs of the five AC circuit loops.

Table 5.4 Parameters of the five-stage IPT system

Parameter	Value/type	Parameter	Value	Parameter	Value
f	100 kHz	C_{BO}	3.542 nF	R_D	800 mΩ
Q_1–Q_4	IRFBA22N50	R_B	700 mΩ	L_{EI}	734.87 μH
D_1–D_4	MBR40200WT	L_{CI}	732.55 μH	L_{EO}	732.46 μH
L_{AI}	735.39 μH	L_{CO}	722.57 μH	C_{EI}	3.447 nF
L_{AO}	726.67 μH	C_{CI}	3.458 nF	C_{EO}	3.456 nF
C_{AI}	3.444 nF	C_{CO}	3.506 nF	R_E	740 mΩ
C_{AO}	3.486 nF	R_C	750 mΩ	M_{AB}	29.14 μH
R_A	760 mΩ	L_{DI}	740.11 μH	M_{BC}	28.95 μH
L_{BI}	720.49 μH	L_{DO}	735.81 μH	M_{CD}	29.32 μH
L_{BO}	715.24 μH	C_{DI}	3.423 nF	M_{DE}	29.43 μH
C_{BI}	3.516 nF	C_{DO}	3.443 nF		

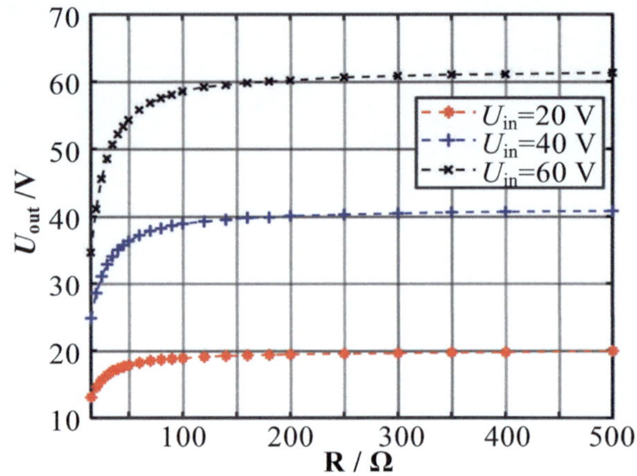

Fig. 5.16 Curves of output voltage varying with load under different input voltages

In order to verify the CV characteristics of the proposed IPT system, multiple sets of experiments are implemented. The load has been changed from 15 Ω to 500 Ω, and the change in output voltage is investigated as shown in Fig. 5.16. Among them, the input voltages are respectively selected as 20, 40 and 60 V.

The result of Fig. 5.16 shows that the system has great CV characteristics. When the load resistance value reaches to a certain level, the value of the output voltage does not change with the load. Finally, the value of the output voltage and the input voltage are nearly equal. Combining the conditions of M_{AB}, M_{BC}, M_{CD}, and M_{DE} in Table 5.4, the conclusion in (5.28) is verified.

Figure 5.17 shows that the five-stage IPT system based on domino cylindrical solenoid coupler compensated respectively in layers can improve efficiency and extend wireless power transfer distance of the system. Among them, in order to better show the change trend of efficiency, the logarithmic abscissa is applied. When the load resistance value changes in a wide range, efficiency of the 2.5 m long-range IPT system exceeding 60% is achieved.

It can be seen from the analysis of the experimental results obtained in Fig. 5.17 that the load resistance value changes from 15 to 500 Ω, the highest DC-DC efficiency of the system can reach 66.7% when $U_{in} = 20$ V and $R = 30$ Ω. The power could be transferred up to 77.8 W and maintain a high efficiency of 62% when $U_{in} = 60$ V and $R = 30$ Ω. Under the above two conditions, the values of voltage, current, power and efficiency on the DC side, and the waveforms of voltage and current on the AC side are shown in Figs. 5.18 and 5.19.

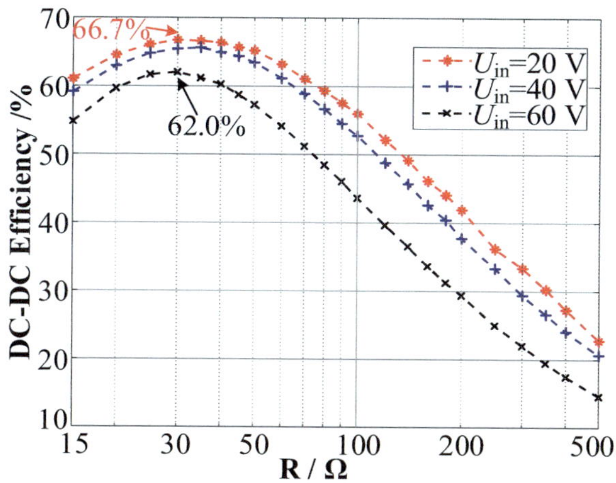

Fig. 5.17 Curves of efficiency varying with load under different input voltages

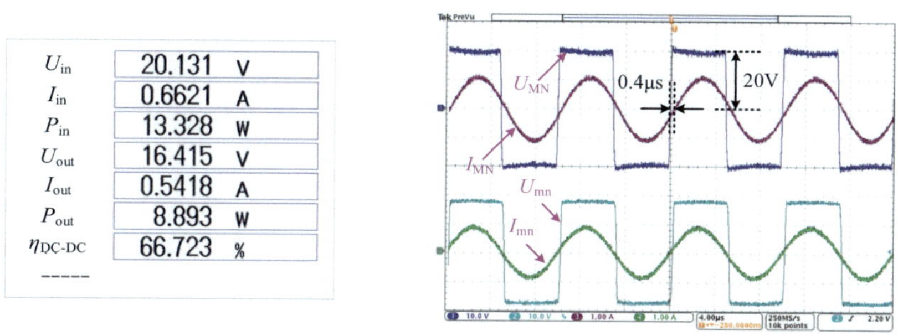

(a) Power and efficiency on the DC side. (b) Voltage and current waveforms on the AC side.

Fig. 5.18 Voltage, current, power and efficiency when $U_{in} = 20$ V and $R = 30\ \Omega$

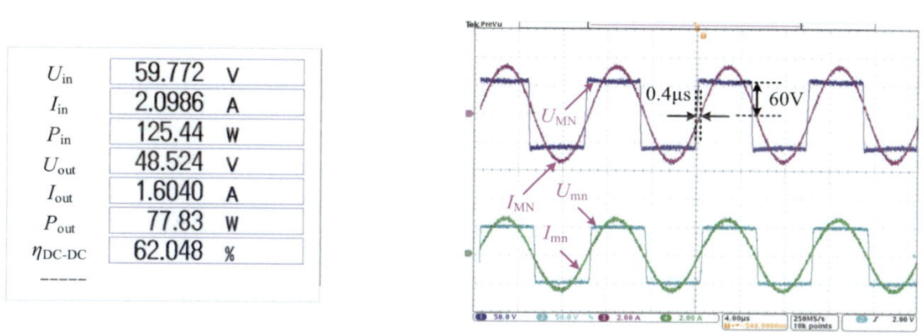

(a) Power and efficiency on the DC side. (b) Voltage and current waveforms on the AC side.

Fig. 5.19 Voltage, current, power and efficiency when $U_{in} = 60$ V and $R = 30\ \Omega$

Figure 5.18 corresponds to the maximum point of the system efficiency, while Fig. 5.19 corresponds to the case of transferring larger power. Phase of U_{MN}, I_{MN}, U_{mn}, and I_{mn} are basically the same, and there are no obvious harmonics as shown in the waveforms, indicating the rationality of the circuit structure. The phase of the input voltage U_{MN} is slightly ahead of the current I_{MN}, indicating that the switch tubes of the inverter have achieved soft switching.

5.4 Conclusion

This chapter proposes a 2.5 m five-stage wireless power transmission system based on double-layer three-dimensional solenoids under distributed series compensation.

A design method of a double-layer three-dimensional solenoid is proposed, so that the self-inductance of the inner and outer coils of the single magnetic structure is basically the same.

In view of the characteristics of the above magnetic coupling structure, the characteristics of four sets of connection and compensation methods for double-layer three-dimensional solenoids are proposed, analyzed and compared, and the superior characteristics of the distributed series compensation method are explained. Compared with the two parallel compensation schemes, this method effectively reduces the main circuit current of the system and significantly reduces the losses of the inverter module, rectifier module and compensation capacitor. Compared with the centralized series compensation method, this method can significantly reduce the loss caused by the voltage difference between the inner and outer coils due to stray capacitance and parasitic resistance. Next, the above conclusion was verified through four sets of experimental comparisons.

A five-stage wireless power transmission system based on double-layer three-dimensional solenoids under distributed series compensation was constructed, and modeling and analysis were conducted to establish an experimental platform. Through experimental analysis, the constant voltage output characteristics of the system are verified. Experimental results show that the maximum efficiency of the system can reach 66.7%. When the power reaches 77.8 W, the system efficiency can still reach 62%.

References

1. Dimitrakakis GS, Tatakis EC, Rikos EJ (2008) A semiempirical model to determine HF copper losses in magnetic components with nonlayered coils. IEEE Trans Power Electron 23(6):2719–2728

2. Podoltsev AD, Kucheryavaya IN, Lebedev BB (2003) Analysis of effective resistance and eddy-current losses in multiturn winding of high-frequency magnetic components. IEEE Trans Magn 39(1):539–548
3. Reinert J, Brockmeyer A, De Doncker RWAA (2001) Calculation of losses in ferro-and ferrimagnetic materials based on the modified Steinmetz equation. IEEE Trans Ind Appl 37(4):1055–1106

Chapter 6
Reactive Power Control of Rectifier Bridge in Long Range IPT with Light Load

Abstract Rectifier circuits are a necessary part of long range IPT systems, and reducing reactive power caused by rectifiers can help improve system efficiency. Bridge rectifier circuit is the most widely used, according to the different filter circuits can be divided into C filter rectifier and LC filter rectifier. The LEACI of the C-filter rectifier is inductive when the load resistance value is large, and the LEACI of the LC-filter rectifier is capacitive when the load resistance value is small. The reactive power caused by the reactance component increases system losses, destroys the resonant state of the compensation network, and increases the difficulty of closed-loop control and system modeling. There are errors in the calculation and analysis. This chapter proposes an analysis method for bridge rectifier LEACI, and proposes an improved rectifier circuit structure and parameter design method to reduce reactance component of LEACI, which is helpful to increase the operating range of the load and improve the system efficiency.

Keyword Long-range inductive power transfer · Reactive power · Circuits stray

6.1 Research and Improvement Design of Reactive Power Characteristics of C-Filter Bridge Rectification Circuit

The C-filtered full-bridge rectifier circuit is shown in Fig. 6.1, where u_{re} and i_{re} are the input voltage and current of the rectifier. Z_{re} is the equivalent AC impedance of the load, and its equation is shown in the Eq. (6.1). U_{re1} is the fundamental of the input voltage and I_{re1} is the fundamental of the input current.

$$Z_{\mathrm{re}} = \frac{U_{\mathrm{re}}}{I_{\mathrm{re}}} \tag{6.1}$$

The harmonics introduced by the rectifier circuit affect the fundamental impedance characteristics of the load. As the load resistance increases, the harmonic distortion rate of the rectifier input current increases, causing the current zero-crossing point to be ahead of the zero-crossing point of the fundamental current. The input current

Y. Wang et al., *Long Range Inductive Power Transfer*, CPSS Power Electronics Series,
https://doi.org/10.1007/978-981-96-7541-8_6

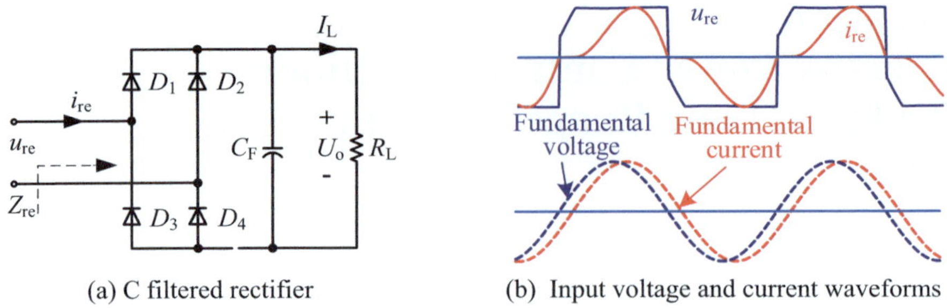

Fig. 6.1 C-filter rectifier and its input voltage and current waveforms

of the rectifier is distorted and lags behind the input voltage. Within a certain load range, the rectifier is in CCM, and its input voltage is a square wave. When the load resistance is further increased, the rectifier will enter the DCM, its input voltage will also be distorted, and the input current will further lag behind the input voltage. A schematic diagram of the input voltage and current waveforms of the rectifier is shown in Fig. 4.1b. LEACI contains an inductive component, which causes reduced system efficiency and increased output fluctuations, which requires in-depth study.

6.1.1 Analysis of Reactive Power Characteristics of Traditional C-Filter Rectifier

To simplify the analysis, the diodes of the rectifier use an ideal model, and the normalized input voltage and current waveforms of the C-filter rectifier in CCM are shown in Fig. 6.2. The input voltage is a square wave, and the input current is an approximate sinusoidal waveform superimposed with multiple harmonics. The current waveform is decomposed into fundamental currents and harmonic currents. The fundamental current lags behind the fundamental voltage, and the load equivalent AC impedance is weakly inductive.

The formula (6.2) is obtained by the conservation of power and the current at the moment of phase conversion being zero. θ_{re} is the load equivalent AC impedance angle, I_{har0} is the harmonic current value at the moment of phase conversion, I_{re1_peak} is the peak of the fundamental current, and U_o is the load voltage. Harmonic current i_{har} is shown in the Eq. (6.3), where ω_1 is the fundamental angle frequency and Z_{2m+1} is the system output impedance at the 2m + 1 harmonic frequency. The modulus of the formula (6.3) and (6.4) the available LEACI $|Z_{re}|$ and the angle θ_{re} are shown in the Eq. (6.4). From Eq. (6.4), θ_{re} is positively correlated with the load resistance and negatively correlated with the harmonic output impedance in CCM. $|Z_{re}|$ is proportional to $\cos(\theta_{re})$, and when $\theta_{re} = 0$, Z_{re} can be equivalent to pure resistance. From Eq. (6.2), the relationship between the input and output currents of the rectifier is shown in the Eq. (6.5). When the output current is the same, the higher the value of

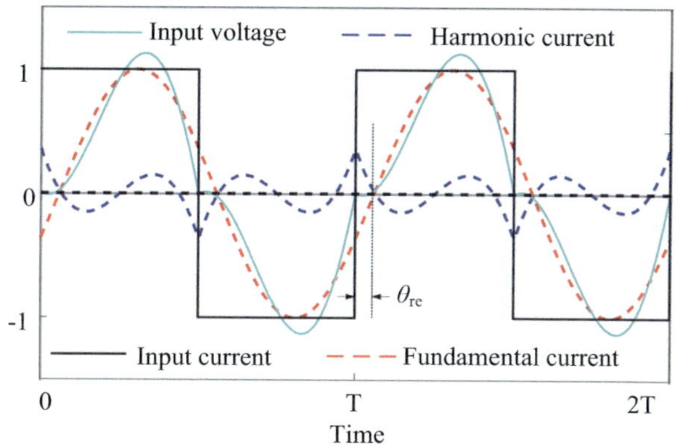

Fig. 6.2 Input voltage and current of C-filtered rectifier in continuous mode

θ_{re}, the greater the input current of the rectifier, which increases the system losses and thus causes the system efficiency to decrease.

$$\begin{cases} I_{har0} + I_{re1\_peak}\sin(\theta_{re}) = 0 \\ \frac{2}{\pi} I_{re1\_peak} U_o \cos(\theta_{re}) = \frac{U_o^2}{R_L} \end{cases} \tag{6.2}$$

$$i_{har} = \sum_{m=1}^{+\infty} \frac{-4U_o}{\pi(2m+1)Z_{2m+1}} \cos((2m+1)\omega_1 t) \tag{6.3}$$

$$\begin{cases} |Z_{re}| = \frac{8R_L}{\pi^2}\cos(\theta_{re}) \\ \theta_{re} = \arctan \sum_{m=1}^{+\infty} \frac{-8R_L}{\pi^2(2m+1)|Z_{2m+1}|} \end{cases} \tag{6.4}$$

$$\frac{2}{\pi} I_{re1\_peak}\cos(\theta_{re}) = I_o \tag{6.5}$$

From the Eq. (6.3), the slope of the harmonic current at the initial moment of phase conversion can be obtained as shown in Eq. (6.6), and L_Z is the equivalent inductance value corresponding to the high-frequency output reactance of the system. Under critical mode conditions, input and output parameters of the rectifier not only satisfy Eq. (6.2), but also meet the condition that the slope of the input current at the initial moment of phase conversion is equal to zero, then get Eq. (6.7).

$$\left.\frac{di_{har}}{dt}\right|_{t=0+} = \frac{U_o}{L_Z} \tag{6.6}$$

$$\left.\frac{di_{har}}{dt}\right|_{t=0+} = \left.\frac{di_1}{dt}\right|_{t=0+} = -\omega_1 I_{re1\_peak}\cos(\theta_{re}) \tag{6.7}$$

From (6.2), (6.6) and (6.7), the critical load resistance R_{LC} is obtained:

$$R_{\mathrm{LC}} = \frac{\pi\omega_1 L_z}{2} \tag{6.8}$$

Taking the S/S compensation topology as an example, Fig. 6.3 shows the IPT system with S/S compensation topology. C_{P1} is the primary series compensation capacitor and C_{S1} is the secondary series compensation capacitor. The circuit parameters are shown in Table 6.1, and the compensation parameters satisfy Eq. (6.9). From Eq. (6.9), the critical resistance of the system is 40.2 Ω, and when the load resistance is less than the critical value, the rectifier is in CCM mode. Under CCM conditions, the calculation results of θ_{re} varying with load are shown in Fig. 6.4a. θ_{re} is basically proportional to the R_{L}, and θ_{re} in critical mode is 18.2°. The input–output current ratio of the rectifier is $\varepsilon_{\mathrm{Ire\_Io}}$ as shown in Fig. 6.4b. As the load increases, $\varepsilon_{\mathrm{Ire\_Io}}$ increases, and the rectifier transmits more reactive power.

$$\begin{cases} C_{\mathrm{P1}} = \frac{1}{\omega_1 L_{\mathrm{P}}} \\ C_{\mathrm{S1}} = \frac{1}{\omega_1 L_{\mathrm{S}}} \end{cases} \tag{6.9}$$

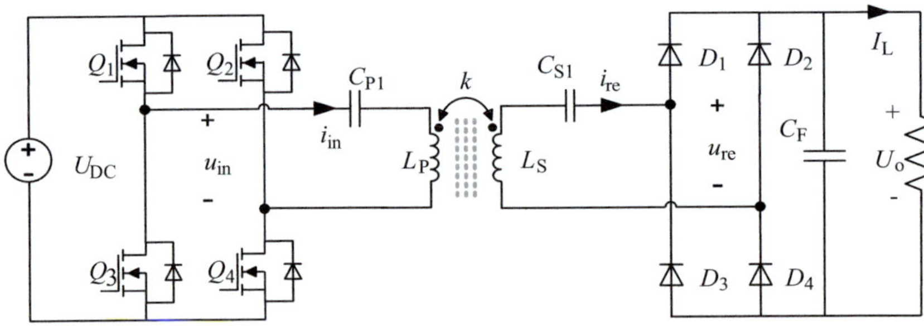

Fig. 6.3 IPT system with S/S compensation

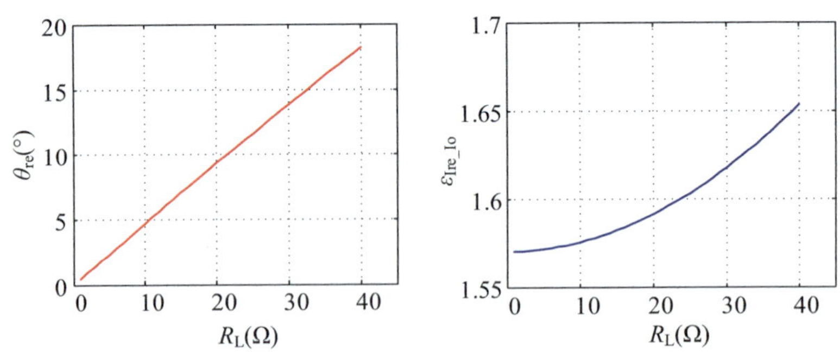

(a) $\varepsilon_{\mathrm{Ire\_Io}}$ versus R_{L} (b) $\varepsilon_{\mathrm{Ire\_Io}}$ versus R_{L}

Fig. 6.4 θ_{re} and $\varepsilon_{\mathrm{Ire\_Io}}$ versus R_{L} in C-filter rectifier

6.1.2 Analysis of Reactive Power Characteristics of Improved C-Filter Rectifier

In order to solve the problem of inductivity of LEACI caused by C-filter rectifier under large load resistance conditions, an improved C-filter rectifier is proposed, as shown in Fig. 6.5a. A capacitor C_{re} is connected in parallel in front of the rectifier, and C_{re} needs to be reverse-charged at the time of phase conversion.

When the voltage of C_{re} is greater than the output voltage, the diode can only turn on, and the phase shift of the input voltage of the rectifier is backward, which is conducive to maintaining the same phase of voltage and current. A schematic of the input voltage and current of the rectifier is shown in Fig. 6.5b.

There are 4 modes in the proposed rectifier in one cycle T, as shown in Fig. 6.6. Modes I and III charge the capacitor C_{re} in reverse, The traditional rectifier has only two modes as shown in Fig. 6.6b, d. The waveforms of each mode are shown in Fig. 6.7. From Fig. 6.7, there is current in C_{re} only when the conduction diode of the rectifier switches.

Mode I: As shown in Fig. 6.6a, occurs when all of the diodes are off and the current $i_D = 0$, at the time 0–t_1 in Fig. 6.7. In this mode, the input current i_{re} continues through the capacitor C_{re} and charge it, which has the same direction with the positive reference, so $i_{re} > 0$. After the voltage of C_{re} before the rectifier is higher than U_L, the diodes return on. Then the rectifier will enter mode II.

Mode II: Mode II happens when the rectifier conducts at the positive half cycle. After mode I, u_{re} is higher than U_L, diodes D_1 and D_4 turn on while D_2 and D_3 turn off, so $i_{D1} = i_{D4} > 0$. The equivalent circuit can be represented by Fig. 6.6b, and the waveforms are shown in Fig. 4.7, at the time t_1–t_2. This mode only appears after C_{re} charging is completed.

Mode II: The equivalent circuit of mode III is shown in Fig. 6.6c, the waveforms are shown in Fig. 6.7 when $t = t_2$–t_3. Its process and analysis are similar to mode I, so it is omitted here.

Mode IV: In mode IV, the rectifier conducts at the negative half cycle, as shown in Fig. 6.6d, at the time of t_3–t_4 in Fig. 6.7. This mode is symmetrical with mode II.

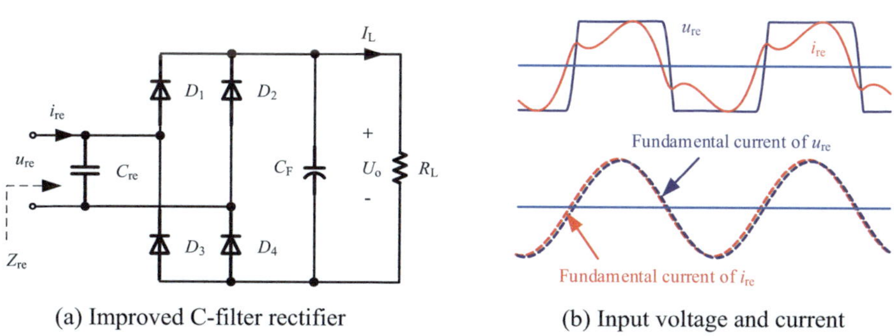

Fig. 6.5 Improved C-filter rectifier and its input voltage and current

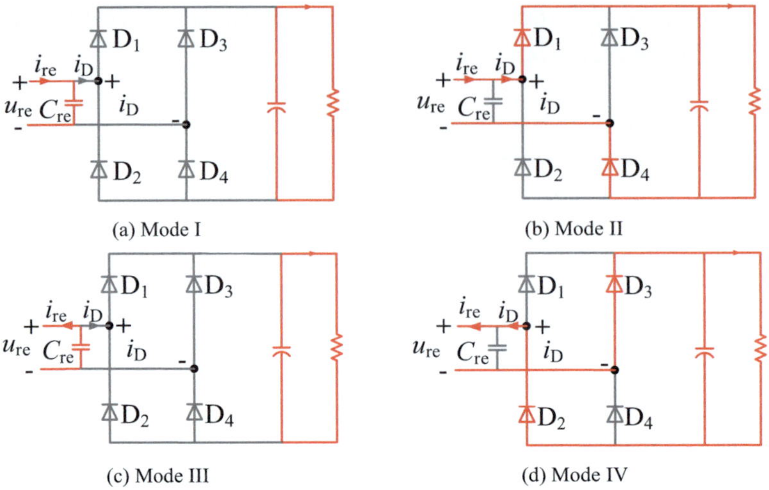

Fig. 6.6 Operating mode of the improved C-filter rectifier

Figure 6.8 is the fundamental wave and harmonic decomposition diagram of the input current and voltage of the improved rectifier. The solid blue line is the input voltage, the blue dashed line is the fundamental wave of the input voltage, the solid red line is the input current, the red dashed line is the fundamental wave of the input current, and the black dashed line is the harmonic of the input current. The phase difference between the voltage fundamental and the current fundamental is θ_{re}, and the phase difference between the current zero crossing and the fundamental voltage zero crossing point is θ_{D1}.

Equation (6.10) can be obtained from the theory of power conservation, zero input current at the time of phase transition, and two-port network. I_{har0_im} is the harmonic current value when the rectifier input current is equal to zero, U_{re1_peak} is the peak fundamental input voltage of the rectifier. U_{in1} and I_{in1} are the fundamental waves of the inverter output voltage and current, respectively. U_{re1} and I_{re1} are the fundamentals of the rectifier input voltage and current, respectively. a_1, b_1, c_1, and d_1 are parameters in the two-port transmission matrix of the compensation network, and a_1 and b_1 satisfy Eq. (6.11). The harmonic current of the improved rectifier is different from the traditional rectifier because the voltage u_{re} is not a square wave and Eq. (6.3) is no longer applicable. The harmonic current calculation formula of the improved rectifier is shown in Eq. (6.12). U_{re_m} is the m-th harmonic voltage of u_{re}. When the C_{re} voltage is reversed, i_{re} charges it, then u_{re} and i_{re} satisfy (6.13). The equivalent load AC impedance of the improved rectifier can be obtained from Eqs. (6.10)–(6.13). However, the equation contains infinite series and nonlinear equations, so Z_{re} cannot be expressed by a concise formula, and this paper uses an iterative solution to calculate Z_{re}.

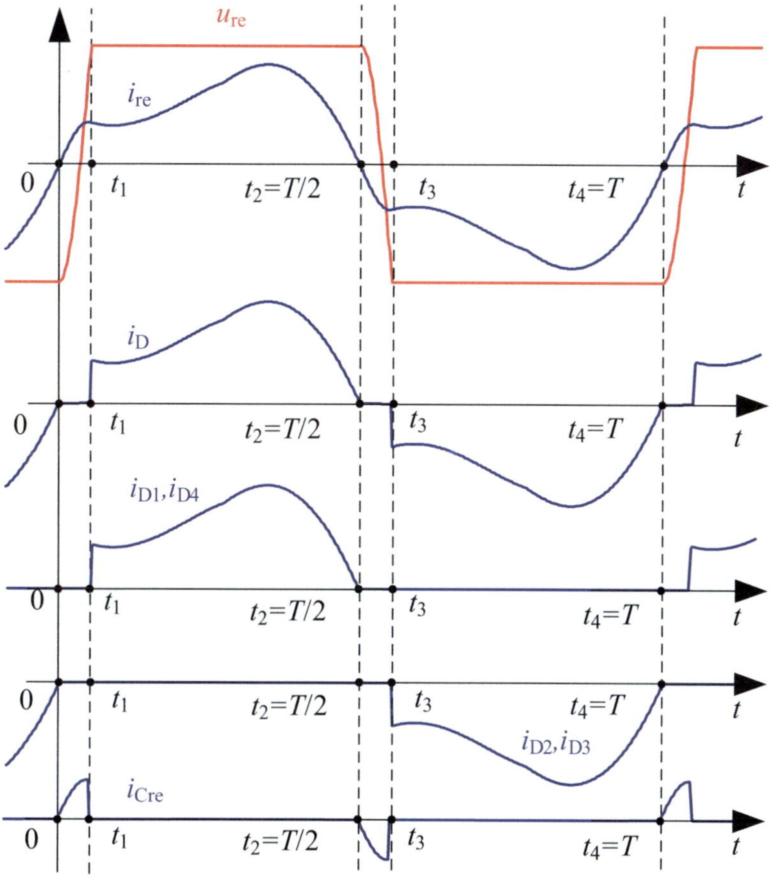

Fig. 6.7 Waveform of the improved C-filter rectifier

$$\begin{cases} I_{har0\_im} + I_{re1\_peak}\sin(\theta_{D1} - \theta_{re}) = 0 \\ \frac{1}{2}U_{re1\_peak}I_{re1\_peak}\cos(\theta_{re}) = \frac{U_o^2}{R_L} \\ U_{in1} = U_{re1}a_1 + I_{re1}b_1 \end{cases} \tag{6.10}$$

$$\begin{bmatrix} U_{in1} \\ I_{in1} \end{bmatrix} = \begin{bmatrix} a_1 & b_1 \\ c_1 & d_1 \end{bmatrix} \begin{bmatrix} U_{re1} \\ I_{re1} \end{bmatrix} \tag{6.11}$$

$$i_{har\_im} = \sum_{m=1}^{+\infty} \frac{-U_{re\_m}}{Z_{2m+1}} \tag{6.12}$$

$$u_{re} = \frac{1}{C_{re}} \int i_{re} \tag{6.13}$$

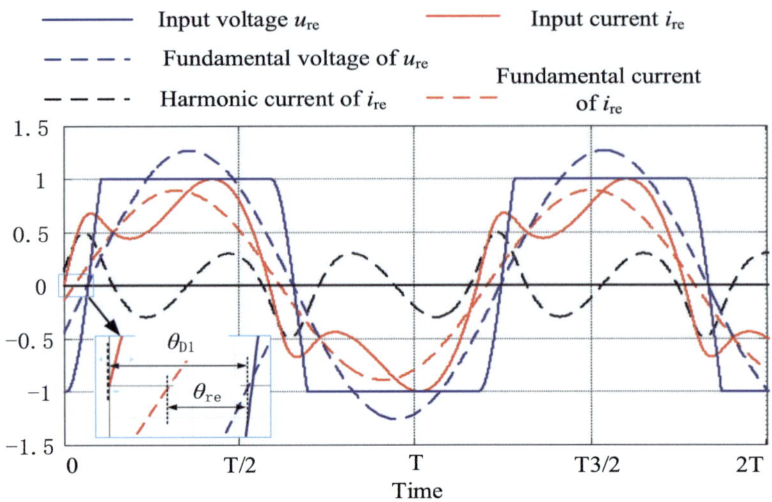

Fig. 6.8 Input voltage and current of the improved C-filter rectifier

Figure 6.9 is the typical schematic of S/S compensation topology with improved rectifier. The load R_L range is 0–100 Ω, and the capacitance C_{re} range is 0.5–4.5 nF. The other circuit parameters are shown in Table 4.1. Solve formulas (6.9)–(6.11) to obtain the curves of the input impedance angle of the rectifier varying with the load under different values of capacitance C_{re}. They are shown in Fig. 6.10a. Figure 6.10b is the curves of the input impedance Z_{re} of the rectifier varying with the load under different values of C_{re}. Under the same load condition, the larger the value of C_{re}, the smaller the θ_{re}. When $C_{re} = 2.5$ nF, the value of θ_{re} is near 0. Compared with conventional rectifiers, the load range of continuous mode is increased, and the approximate resistivity of LEACI is maintained. When the value of C_{re} is large, the LEACI transitions from inductive to capacitive.

$$Z_{in} = \frac{(\omega_1 M)^2}{Z_{re}} \tag{6.14}$$

Figure 6.11 shows the circuit simulation results of Z_{re} and θ_{re}. The simulation results are basically consistent with the calculation results, which proves the effectiveness of the proposed analysis method. The main reason for the subtle error is the harmonic influence of the primary inverter. The compensation capacitance parameter of the S/S topology is calculated by Eq. (6.9), then the relationship between the input impedance Z_{in} and the input impedance Z_{re} of the inverter is as shown in Eq. (6.14), and M is the mutual inductance of the LCT. When Z_{re} is inductive, Z_{in} is capacitive; When Z_{re} is capacitive, Z_{in} is inductive. When the impedance angle of Z_{re} is greater than zero, the impedance angle of Z_{in} will be less than zero, and the inverter cannot implement ZVS. Therefore, the equivalent AC impedance change caused by the rectifier can also affect the implementation of the inverter soft switching.

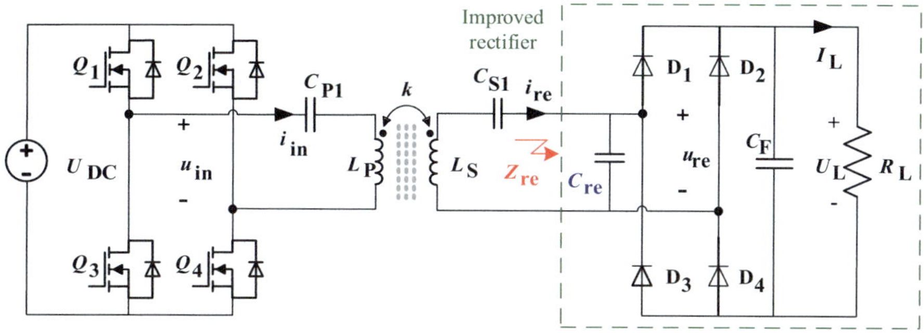

Fig. 6.9 IPT system with S/S compensation and improved rectifier

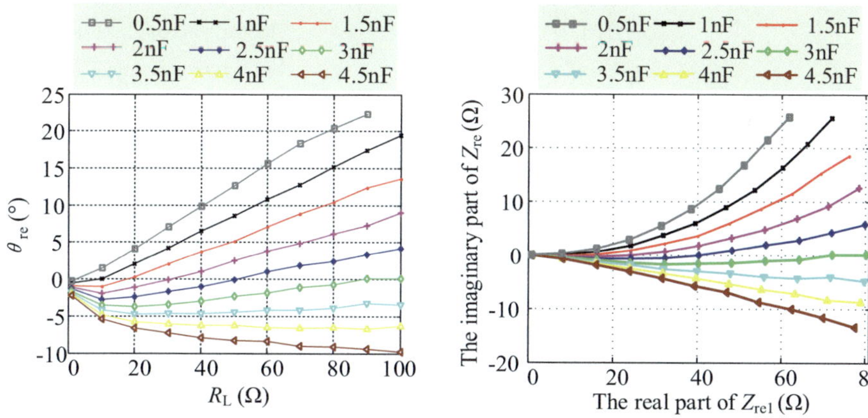

(a) θ_{re} versus R_L and C_{re} (b) Z_{re} versus R_L and C_{re}

Fig. 6.10 Calculation results of θ_{re} and Z_{re} of the improved C-filter rectifier

The above calculation and simulation example is based on the S/S compensation topology as an example. The improved C filter rectifier proposed in this paper is not only suitable for S/S compensation topologies, but also suitable for other compensation topologies that can adopt C-filter bridge rectifier circuits, such as LCC/S [1], LCL/S [2] and LCC/LCC [3] and other topologies. From the above analysis, it can be seen that the high-frequency output impedance of the rectifier is the main factor affecting the reactance component of the LEACI. The smaller the high-frequency output impedance value, the greater the reactance component of the LEACI, and the high-frequency output impedance is basically determined by the high-frequency output inductor L_z. L_z of the LCC/LCC compensation topology is equal to the secondary-side compensation inductor, and L_z of the compensation topology S/S, LCC/S, and LC/S is equal to $L_S(1 - k_2)$. In most applications, the secondary compensation inductance of LCC/LCC is much smaller than the self-inductance of the LCT, so its critical load resistance R_{LC} is relatively small, and there

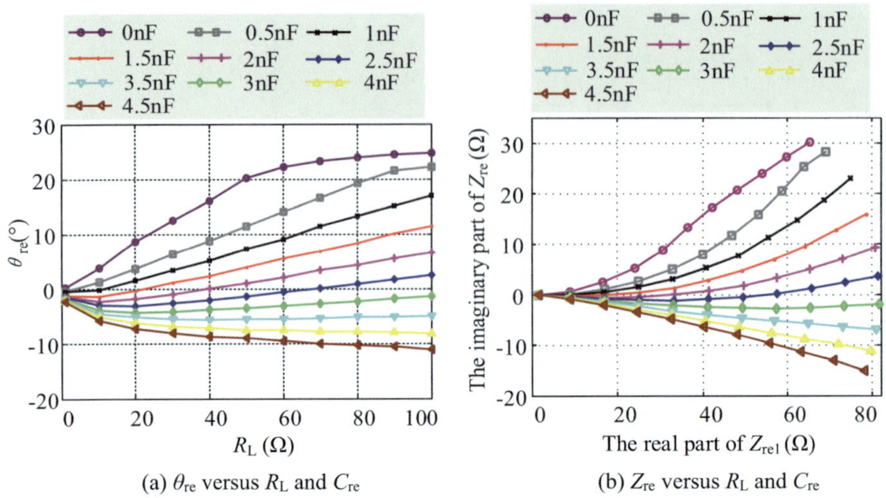

Fig. 6.11 Simulation results of θ_{re} and Z_{re} of the improved C-filter rectifier

are more reactance components under the same load resistance compared to other compensation topologies.

6.1.3 Experimental Verification

In order to verify the performance of the improved C-filter rectifier and the aforementioned theoretical analysis, an IPT system based on S/S compensation topology is constructed, as shown in Fig. 6.12. The load range is 1–100 Ω, the rated power is 1000 W, and the output current is 3.3 A. Based on the above simulation and calculation results, the design C_{re} is 2.5 nF. Compare with an IPT system using a traditional rectifier to verify the performance of the improved rectifier. The LEACI of traditional rectifiers is inductive, the parameters are shown in Table 6.1. The system output impedance Zin will be capacitive. To achieve ZVS, in an IPT system with an improved rectifier, design the primary side compensation capacitor C_{P1} to 60.34 nF. In an IPT system using a conventional rectifier, the primary compensation capacitor, C_{P1}, is adjusted to 62.50 nF.

Figure 6.13a, b are the experimental results of θ_{re} and Z_{re}, respectively. The Z_{re} of conventional rectifiers is inductive with a maximum impedance angle of 25.9° and a minimum impedance angle of 2.1°. In contrast, the reactive weight of Z_{re} in the improved rectifier is smaller, and the impedance angle θ_{re} ranges from –3° to 1.7°, which is close to pure resistivity. There are small errors in the experimental results and simulation, mainly caused by the parasitic parameters and parameter errors of the circuit, and the parasitic capacitance of the diode has the greatest influence.

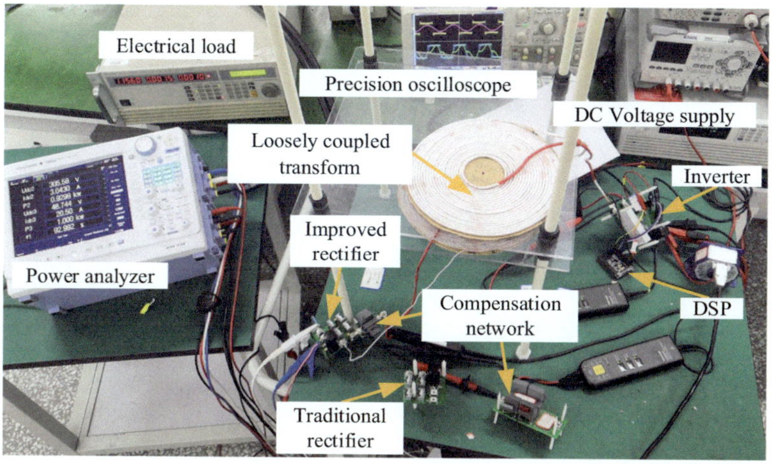

Fig. 6.12 IPT prototype with S/S compensation

Table 6.1 Circuit parameters of the IPT system with S/S compensation

Parameter	Value	Unit	Parameter	Value	Unit	Parameter	Value	Unit
U_{DC}	56	V	L_S	58.5	μH	C_{S1}	59.9	nF
f	85	kHz	k	0.43		I_o	3.4	A
L_P	58.8	μH	C_{P1}	59.6	nF	R_L	1–40.2	Ω

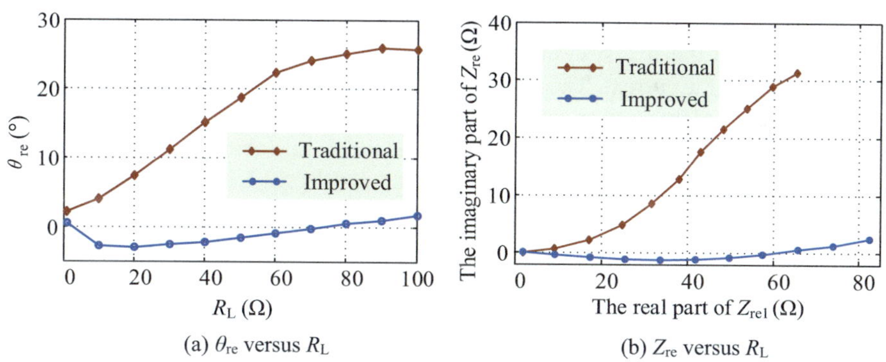

Fig. 6.13 Experimental results of θ_{re} and Z_{re} of the improved C-filter rectifier

The curves of the input–output current ratio ε_{Ire_Io} varying with the load resistance is shown in Fig. 6.14a. For traditional rectifiers, the larger the load resistance, the greater the value of ε_{Ire_Io} up to 1.8. Coil losses are greater, which inevitably reduces system efficiency. For the improvement of rectifier, ε_{Ire_Io} fluctuation with load change is very small, between 1.57–1.63. The results of Fig. 6.14a have a small

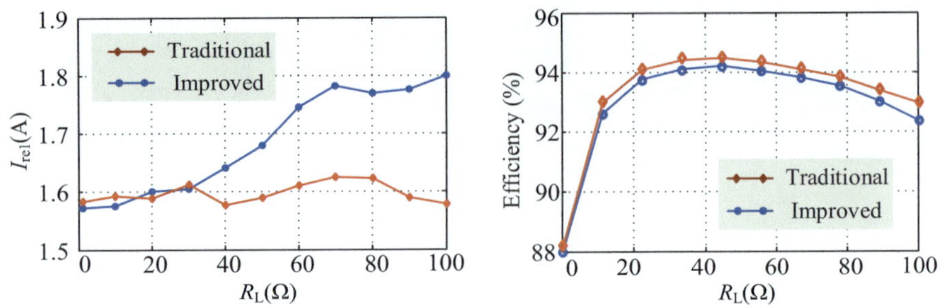

(a) I_{re1} versus R_L (b) System efficiency versus R_L

Fig. 6.14 Experimental results of I_{re1} and system efficiency

error, mainly caused by the measurement error of the current probe. The curve of system efficiency varying with the load resistance is shown in Fig. 6.14b. As the load resistance increases, the system efficiency first increases and then decreases. The efficiency range of systems with traditional rectifiers is 88–94.2%, and the efficiency range of systems with improved rectifiers is 88.2–94.5%. IPT systems with improved rectifiers are more efficient over the full load range, and the larger the load resistance, the greater the efficiency difference between the two systems. Under rated load conditions, the output power of the IPT system with the improved rectifier and the traditional rectifier is 929.9 and 875.1 W, the system efficiency is 93.0 and 92.4%, respectively, and the maximum efficiency difference is 0.6%.

Figure 6.15 is the output voltage/current of the inverter and the input voltage/current of the rectifier when the load resistance is equal to 100 Ω. The inverter output voltage of the IPT system with improved rectifier and traditional rectifier is slightly ahead of the current. The system input impedance is weakly inductive, which facilitates the implementation of ZVS. However, the input voltage of a traditional rectifier oscillates at the moment of phase transition, and the current is limited to zero. The voltage is ahead of the current, which causes the LEACI to be inductive. The input voltage of the improved rectifier does not oscillate because the capacitor C_{re} suppresses voltage oscillation by reverse charging at the time of phase transition. The phase of the input voltage and current is essentially the same, and the rectifier transmits almost no reactive power.

6.2 Research and Improvement Design of Reactive Power Characteristics of LC-Filter Bridge Rectification Circuit

The LC filter full-bridge rectifier circuit is shown in Fig. 6.16a, where u_{re} and i_{re} are the input voltage and current of the rectifier.

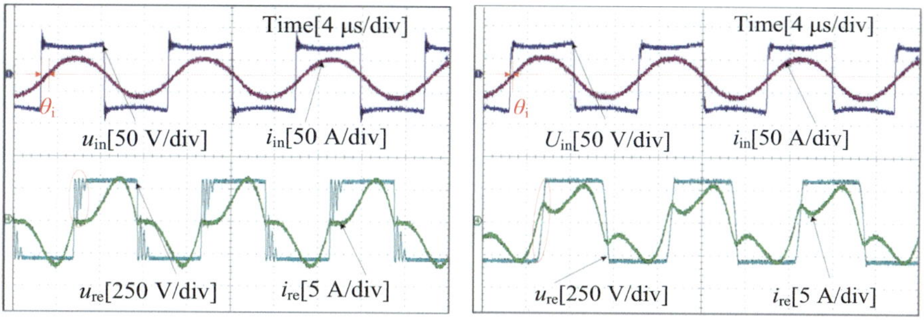

(a) Traditional IPT systems (b) Improved IPT systems

Fig. 6.15 Inverter output voltage/current and rectifier input voltage/current

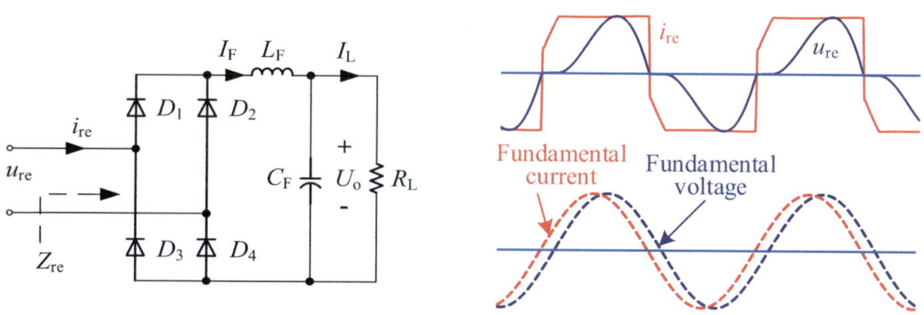

(a) LC filtered rectifier (b) Input voltage and current waveforms

Fig. 6.16 LC-filtered and LC-filtered rectifiers

The DCM of the LC-filter rectifier occurs when the load resistance is relatively small. As the load resistance decreases, the total harmonic distortion of the rectifier input voltage increases. The input voltage of the rectifier is distorted and lagged behind the input current, the LEACI is capacitive, and the input voltage and current waveform diagram is shown in Fig. 6.16b. The LEACI contains a capacitive component, which causes reduced system efficiency and increased output fluctuations. This issue needs to be studied in depth.

6.2.1 Analysis of Reactive Power Characteristics of Traditional LC-Filter Rectifier

To simplify the analysis, it is also assumed that the rectifier diode is the ideal model, the filter inductance is infinite. The normalized input voltage and current waveforms of the LC filter rectifier in continuous mode are shown in Fig. 6.17. The input current of the rectifier is a square wave, and the input voltage is an approximate sinusoidal waveform superimposed with multiple harmonics. This input voltage waveform is decomposed into fundamental voltage and harmonic voltage. The fundamental current is ahead of the fundamental voltage, and the LEACI is capacitive.

The formula (6.15) is obtained by the conservation of power and the current at the moment of phase conversion being zero. U_{har0} is the harmonic voltage value at the moment of phase conversion, $U_{\text{re1\_peak}}$ is the peak of the fundamental voltage, and I_o is the load current. Harmonic voltage u_{har} is shown in the Eq. (6.16). From formula (6.15) and (6.16), $|Z_{\text{re}}|$ and θ_{re} are shown in the Eq. (6.17). From Eq. (6.17), θ_{re} is negatively correlated with the load resistance and positively correlated with the harmonic output impedance in CCM. $|Z_{\text{re}}|$ is inversely proportional to $\cos(\theta_{\text{re}})$, and when $\theta_{\text{re}} = 0$, Z_{re} can be equivalent to pure resistance. From Eq. (6.15), the relationship between the input and output voltages of the rectifier is shown in the Eq. (6.18). On the one hand, when the output voltage is the same, the higher the value of θ_{re}, the greater the input voltage of the rectifier, the greater the current of the capacitor connected in parallel in front of the rectifier, which increases the system losses. On the other hand, I_{re} is added to the current of the parallel capacitor in front of the rectifier, which increases the current of the series branch on the front side. The loss of components on this branch is increased. These two factors will reduce the efficiency of the system.

$$\begin{cases} U_{\text{har0}} + U_{\text{re1\_peak}} \sin(\theta_{\text{re}}) = 0 \\ \frac{2}{\pi} U_{\text{re1\_peak}} I_o \cos(\theta_{\text{re}}) = I_o^2 R_L \end{cases} \tag{6.15}$$

$$u_{\text{har}} = \sum_{m=1}^{+\infty} \frac{4 I_o Z_{2m+1}}{\pi(2m+1)} \cos((2m+1)\omega_1 t) \tag{6.16}$$

$$\begin{cases} |Z_{\text{re}}| = \frac{\pi^2 R_L}{8\cos(\theta_{\text{re}})} \\ \theta_{\text{re}} = \arctan \sum_{m=1}^{+\infty} \frac{-8|Z_{2m+1}|}{\pi^2(2m+1)R_L} \end{cases} \tag{6.17}$$

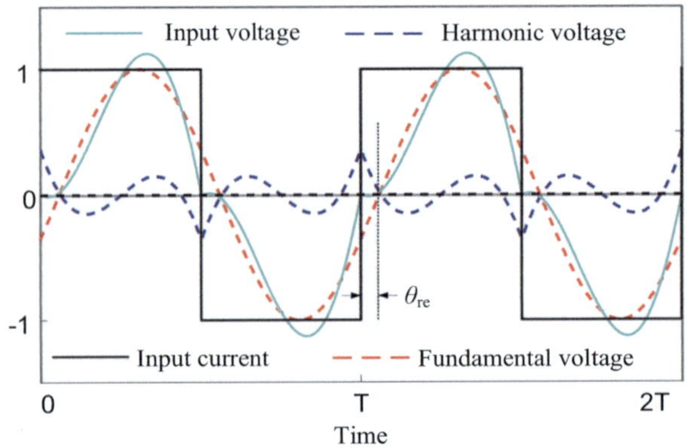

Fig. 6.17 Input voltage and current of LC-filtered rectifier in continuous mode

$$\frac{2}{\pi} U_{\mathrm{re1\_peak}} \cos(\theta_{\mathrm{re}}) = U_{\mathrm{o}} \tag{6.18}$$

From the Eq. (6.16), the slope of the harmonic voltage at the initial moment of phase conversion can be obtained as shown in Eq. (6.19), and C_Z is the equivalent capacitance value corresponding to the high-frequency output reactance of the system. Usually, its value is equal to the value of the parallel capacitor before the LC-filter rectifier. Under critical mode conditions, input and output parameters of the rectifier not only satisfy Eq. (6.15), but also meet the condition that the slope of the input voltage at the initial moment of phase conversion is equal to zero, then get Eq. (6.20).

$$\left.\frac{\mathrm{d}u_{\mathrm{har}}}{\mathrm{d}t}\right|_{\mathrm{t=0+}} = \frac{I_{\mathrm{o}}}{C_{\mathrm{Z}}} \tag{6.19}$$

$$\left.\frac{\mathrm{d}u_{\mathrm{har}}}{\mathrm{d}t}\right|_{\mathrm{t=0+}} = \left.\frac{\mathrm{d}u_1}{\mathrm{d}t}\right|_{\mathrm{t=0+}} = -\omega_1 U_{\mathrm{o\_peak}} \cos(\theta_{\mathrm{re}}) \tag{6.20}$$

From (6.15), (6.19) and (6.20), the critical load resistance R_{LC} is obtained:

$$R_{\mathrm{LC}} = \frac{2}{\pi \omega_1 C_{\mathrm{Z}}} \tag{6.21}$$

Taking the S/SP compensation topology as an example, Fig. 6.18 shows the IPT system with S/SP compensation topology. The circuit parameters are shown in Table 6.2, and the compensation parameters satisfy Eq. (6.22). From Eq. (6.22), the critical resistance of the system is 10.9 Ω.

$$\begin{cases} C_{\mathrm{P1}} = \dfrac{1}{\omega_1^2 L_{\mathrm{P}}(1-k)} \\ C_{\mathrm{S1}} = \dfrac{1}{\omega_1^2 L_{\mathrm{S}}(1-k)} \\ C_{\mathrm{S2}} = \dfrac{1}{\omega_1^2 k \sqrt{L_{\mathrm{P}} L_{\mathrm{S}}}} \end{cases} \tag{6.22}$$

When the load resistance is greater than the critical value, the rectifier is in CCM mode. Under CCM conditions, the calculation results of θ_{re} varying with load are shown in Fig. 6.19a. θ_{re} increases with the increase of the load resistance, and θ_{re} in critical mode is $-16.8°$. The input–output voltage ratio of the rectifier is $\varepsilon_{\mathrm{ure\_uo}}$ as shown in Fig. 6.19b. As the load decreases, $\varepsilon_{\mathrm{ure\_uo}}$ increases, and the rectifier transmits more reactive power.

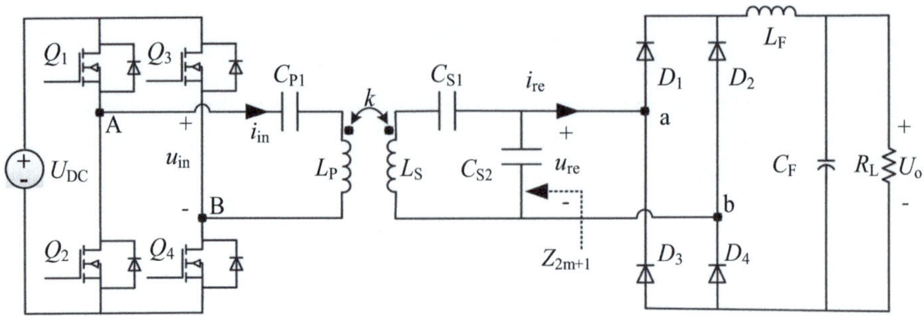

Fig. 6.18 IPT system with traditional S/SP compensation

Table 6.2 Circuit parameters of the IPT system with traditional S/SP compensation

Parameter	Value	Unit	Parameter	Value	Unit
U_{DC}	60	V	C_{P1}	28.1	nF
f	85	kHz	C_{S1}	27.2	nF
L_P	155.7	μH	C_{S2}	108.9	nF
L_S	161.0	μH	R_L	10.9–80	Ω
k	0.20		U_o	49.3	V

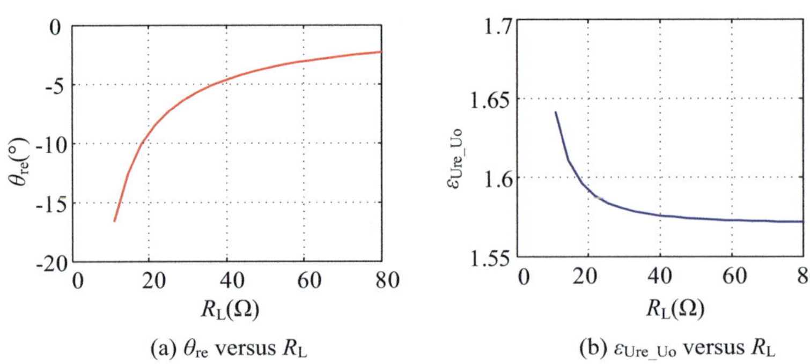

Fig. 6.19 θ_{re} and ε_{Ure_Uo} versus R_L in LC-filter rectifier

6.2.2 Analysis of Reactive Power Characteristics of Improved LC-Filter Rectifier

In order to solve the problem of capacitance of LEACI caused by LC-filter rectifier under small load resistance conditions, an improved LC-filter rectifier is proposed, as shown in Fig. 6.20a. An inductor L_{re} is connected in series in front of the rectifier, and L_{re} needs to be reverse-charged at the time of phase conversion.

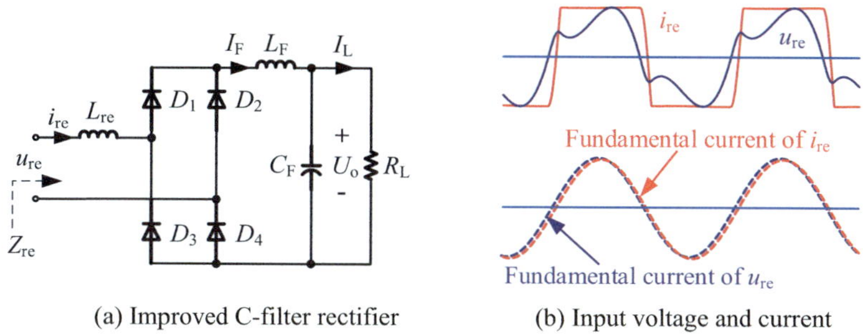

Fig. 6.20 Improved LC-filter rectifier and its input voltage and current

When the current of L_{re} is greater than the current of the filtered inductor L_F, the diode can only turn on, and the phase shift of the input current of the rectifier is backward, which is conducive to maintaining the same phase of voltage and current. A schematic of the input voltage and current of the rectifier is shown in Fig. 6.20b.

There are 6 modes in the proposed LC-filter rectifier in one cycle T, as shown in Fig. 6.21. Modes I, III, IV and VI charge the capacitor L_{re} in reverse, and Modes II and IV charge the load. The traditional rectifier has only two modes as shown in Fig. 6.21b, e. The waveforms of each mode are shown in Fig. 6.22. From Fig. 6.22, there is voltage in L_{re} only when the conduction diode of the rectifier switches. According to symmetry, only half of the modes need to be analyzed.

Mode I: This mode is shown in Fig. 6.21a, and the waveforms of current and voltage are from 0 to t_1 in Fig. 6.22. As shown, the current i_{D1} and i_{D4} increases but i_{D2} and i_{D3} decreases, all of the diodes are on until i_{D2} (i_{D3}) $= 0$. After that, diodes

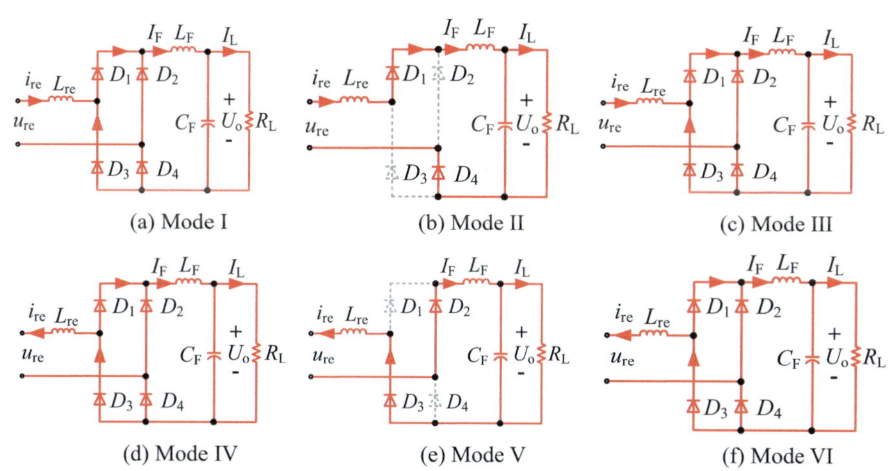

Fig. 6.21 Operating mode of the improved LC-filter rectifier

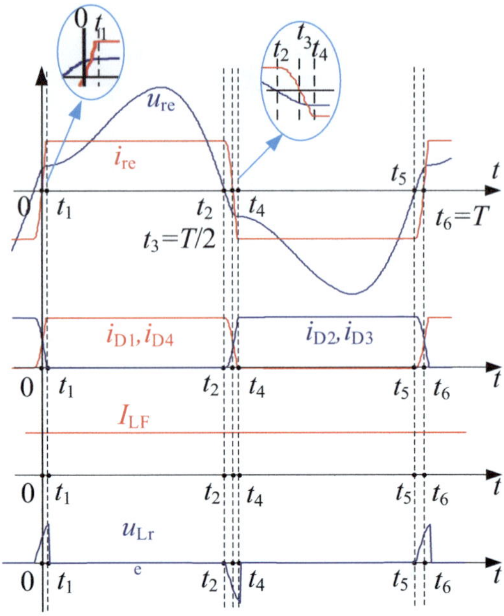

Fig. 6.22 Waveform of the improved LC-filter rectifier

D_2 and D_3 will turn off and the improved rectifier will enter *Mode* II. Note that the current $i_{re} > 0$ is an important condition to distinguish *Mode* I from *Mode* VI.

Mode II: When i_{D1} and i_{D4} stopped growing and reached a stable value i_{LF} by the filter inductor L_F, the improved rectifier enters *Mode* II, as shown in Fig. 6.21b. During the time t_1–t_2, diodes D_1 and D_4 are on while diodes D_2 and D_3 are off. This state takes up most of the positive half cycle.

Mode III: The equivalent circuit is shown in Fig. 6.21c, the waveforms are shown in Fig. 6.22 from t_2–t_3. When $t = t_2$, the voltage u_{re} changes from positive to negative and crosses zero. Thus, negative voltage is applied to inductor L_{re} and current i_{re} decreases from the stable value i_{LF}. In the process, i_{D2} and i_{D3} increases while i_{D1} and i_{D4} decreases. Until $i_{re} = i_{D1} - i_{D2}$ (t = t3). *Mode* II ends and the improved rectifier is ready to *Mode* IV.

Mode IV, *Mode* V and *Mode* VI: They are symmetrical to *Mode* I, *Mode* II and *Mode* II, respectively.

Figure 6.23 is the fundamental wave and harmonic decomposition diagram of the input current and voltage of the improved rectifier. The solid blue line is the input voltage, the blue dashed line is the fundamental wave of the input voltage, the solid red line is the input current, the red dashed line is the fundamental wave of the input current, and the black dashed line is the harmonic of the input voltage. The phase difference between the voltage fundamental and the current fundamental is θ_{re}, and the phase difference between the voltage zero crossing and the fundamental current zero crossing point is θ_{D2}.

Equation (6.23) can be obtained from the theory of power conservation, zero input voltage at the time of phase transition, and two-port network. U_{har0_im} is the harmonic

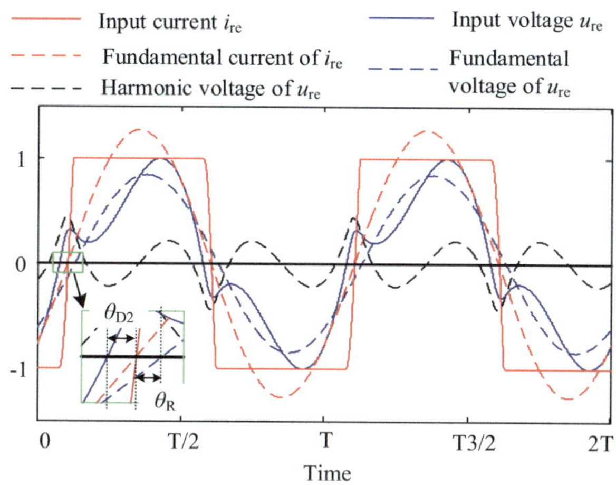

Fig. 6.23 Input voltage and current of the improved LC-filter rectifier

voltage value when the rectifier input voltage is equal to zero. U_{re1_peak}, U_{in1}, I_{in1}, U_{re1}, a_1 and b_1 have the same meanings as defined in the previous section, and a_1 and b_1 are equally satisfied (6.11). In improved rectifiers, i_{re} is not a square wave and Eq. (6.16) is no longer applicable. The harmonic voltage calculation formula of the improved rectifier is shown in Eq. (6.24). I_{re_m} is the m-th harmonic current of I_{re}. When the L_{re} voltage is reversed, u_{re} is equal to the voltage of L_{re}, then u_{re} and i_{re} satisfy (6.25). Z_{re} and θ_{re} of the improved rectifier can be iteratively obtained from Eqs. (6.10), (6.23)–(6.25).

$$\begin{cases} U_{har0\_im} + U_{re1\_peak}\sin(\theta_{D2} - \theta_{re}) = 0 \\ \frac{1}{2} I_{re1\_peak} U_{re1\_peak}\cos(\theta_{re}) = \frac{U_o^2}{R_L} \\ U_{in1} = U_{re1}a_1 + I_{re1}b_1 \end{cases} \tag{6.23}$$

$$U_{har\_im} = \sum_{m=1}^{+\infty} -I_{re\_m} Z_{2m+1} \tag{6.24}$$

$$i_{re} = \frac{1}{L_{re}} \int u_{re} \tag{6.25}$$

Figure 6.24 is the typical schematic of S/SP compensation topology with improved rectifier. The load R_L range is 7–80 Ω, and the inductance L_{re} range is 0.1–1.5 nF. The other circuit parameters are shown in Table 6.2. The curves of θ_{re} varying with the R_L under different values of L_{re} are shown in Fig. 6.25a. Figure 6.25b is the curves of Z_{re} varying with R_L. When the value of L_{re} is relatively small, θ_{re} increases with the increase of load resistance. When the value of L_{re} is relatively large, θ_{re} decreases with the increase of load resistance. In the intermediate transition phase, θ_{re} changes very little with the increase of the load resistance and is around 0°.

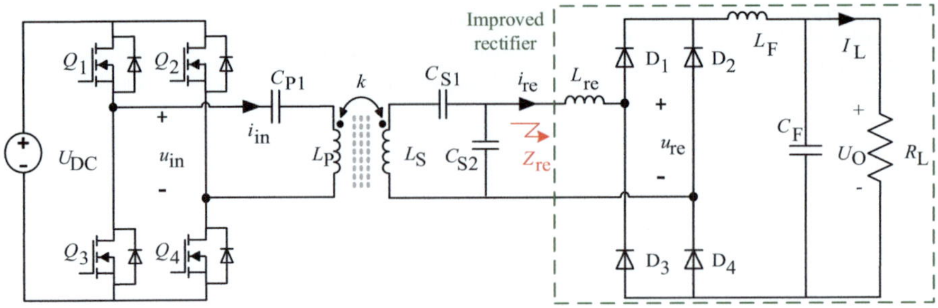

Fig. 6.24 IPT system with traditional S/SP compensation and improved rectifier

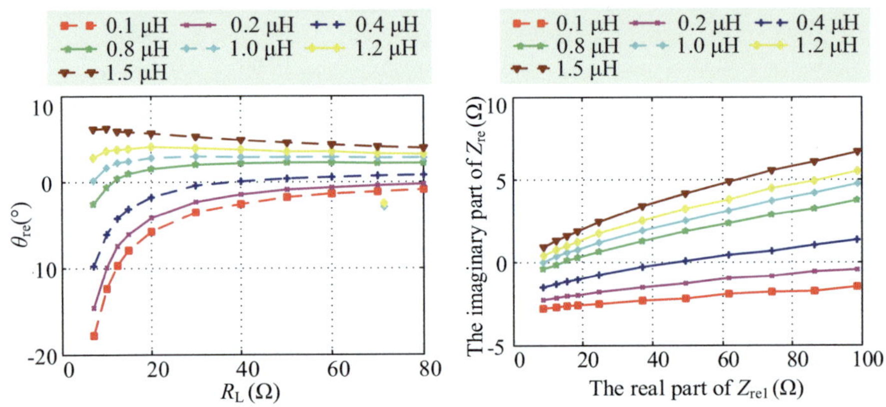

(a) $\theta_{\rm re}$ versus $R_{\rm L}$ and $L_{\rm re}$ (b) $Z_{\rm re}$ versus $R_{\rm L}$ and $L_{\rm re}$

Fig. 6.25 Calculation results of $\theta_{\rm re}$ and $Z_{\rm re}$ of the improved LC-filter rectifier

Figure 6.26 shows the circuit simulation results of $Z_{\rm re}$ and $\theta_{\rm re}$. The simulation results are basically consistent with the calculation results, which proves the effectiveness of the proposed analysis method. The main reason for the subtle error is the harmonic influence of the primary inverter and the current fluctuations of the filtered inductor.

The compensation capacitance parameter of the traditional S/SP topology is calculated by Eq. (6.22), its analysis diagram is shown in Fig. 6.27. $L_{\rm PL}$ and $L_{\rm SL}$ are primary and secondary leakage inductions, respectively. $L_{\rm M}$ is the excitation inductor. $C_{\rm S1}'$, $L_{\rm SL}'$ and $C_{\rm S2}'$ are the calculated values of $C_{\rm S1}$, $L_{\rm SL}$, and $C_{\rm S2}$ from the secondary side to the primary side, respectively. $C_{\rm P1}$ resonates in series with $L_{\rm PL}$, $C_{\rm S1}'$ resonates in series with $L_{\rm SL}'$, $C_{\rm S2}'$ resonates in series with $L_{\rm M}$, and the system fundamental input impedance $Z_{\rm in1} = Z_{\rm re1}$. In IPT systems with traditional LC-filter rectifiers, $Z_{\rm re1}$ is capacitive, so $Z_{\rm in1}$ is also capacitive. It not only causes the inverter to output reactive power, but also affects the realization of ZVS. In IPT system with improved LC-filter

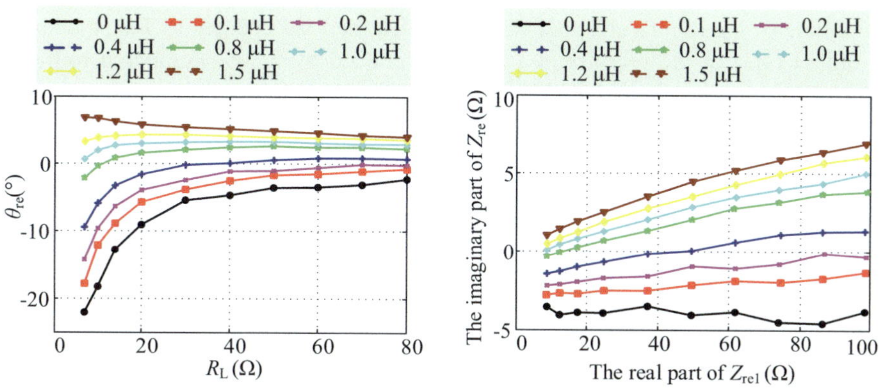

(a) θ_{re} versus R_L and L_{re} (b) Z_{re} versus R_L and L_{re}

Fig. 6.26 Simulation results of θ_{re} and Z_{re} of the improved LC-filter rectifier

rectifier, Z_{re1} can be designed to be weakly inductive, so Z_{in1} is also weakly insensitive, which not only reduces the reactive power, but also facilitates the realization of ZVS.

The vector relationship between the rectifier input current ire, the secondary parallel capacitor current i_{CS2}, and the secondary coil current i_{LS} is shown in Fig. 6.28. i_{LS_im} and i_{LS_tra} are the secondary coil currents of the system with improved and traditional LC-filter rectifier, respectively. i_{CS2_im} and i_{CS2_tra} are improved and secondary parallel capacitor currents of the system with improved and traditional LC-filter rectifier, respectively. In the traditional LC-filter rectifier system, Z_{re} is capacitive, and the phase difference $\theta_{re\_CS2\_tra}$ between i_{re} and i_{CS2_tra} is less than 90°. In the systems with improved LC-filter rectifier, i_{CS2_im} are smaller, and the reactance component of Z_{re} is even inductive. $\theta_{re\_CS2\_im}$ is greater than $\theta_{re\_CS2\_tra}$. When L_{re} equals 0 and 1.2 μH, the ratio ε_{Ure_Uo} of the input and output voltages of the rectifier is shown in Fig. 6.29a. When L_{re} equals 1.2 μH, the rectifier outputs the same voltage and the

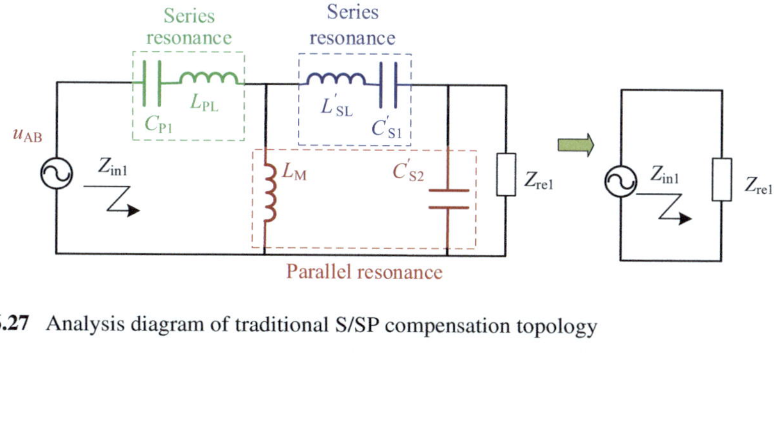

Fig. 6.27 Analysis diagram of traditional S/SP compensation topology

amplitude of the input voltage is smaller. When the load is equal to 7 Ω, the input-to-output ratio of the rectifier at 1.2 μH is reduced by 5.76% compared to 0 μH. When L_{re} is equal to 0 and 1.2 μH, the ratio εILS_Uo of the secondary coil current to output voltage is shown in Fig. 6.29b. When Lre is equal to 1.2 μH, the current of the secondary coil with the same voltage output is smaller. When the load is equal to 7 Ω, the coil current to output voltage ratio at 1.2 μH is reduced by 15.94% compared to 0 μH. Based on the above analysis, compared with the close pure resistivity, the design Z_{re} is weakly inductive, which is conducive to reducing the secondary coil current. Therefore, the system efficiency is improved.

The above calculation and simulation example is based on the S/SP compensation topology. The proposed improved LC-filter rectifier is not only suitable for S/SP compensation topologies, but also for topologies with secondary parallel compensation capacitors before rectifiers such as S/P [4] and S/CLC [5]. The high-frequency output impedance of the rectifier is the main factor affecting reactance component of the LEACI, which is the same as the C-filter rectifier. The larger the high-frequency output impedance, the greater the reactance component of the LEACI, and the high-frequency output impedance is basically determined by the high-frequency output capacitor C_z.

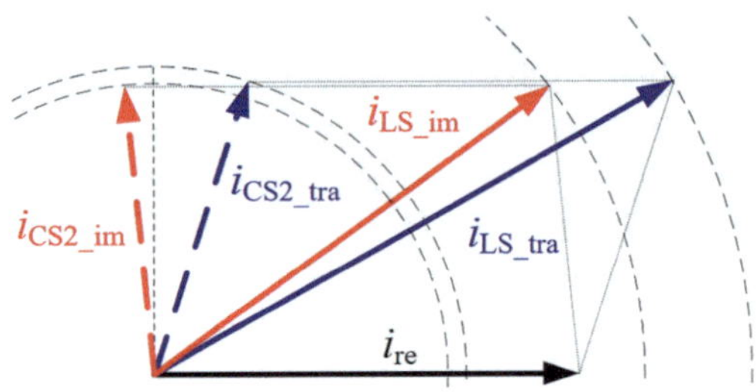

Fig. 6.28 Vectors of i_{re}, i_{CS2} and i_{LS}

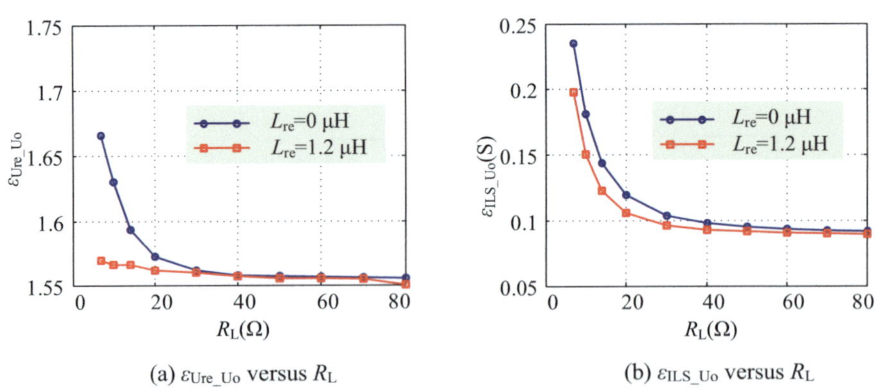

Fig. 6.29 ε_{Ure_Uo} and ε_{ILS_Uo} versus R_L when L_{re} is equal to 0 and 1.2 μH

6.2.3 Experimental Verification

Two experimental prototypes based on the traditional S/SP compensation topology were established to verify the above analysis of LC-filter rectifier. In the system with traditional LC-filter rectifier, where $L_{re} = 0$ μH, Z_{re}, is capacitive. To achieve ZVS, increase the value of C_{P1} to 30.0 nF so that Eq. (6.26) are satisfied. In the system with improved LC-filter rectifier, where $L_{re} = 1.2$ μH, Z_{re}, is weakly inductive. Design the compensation parameters according to Eq. (6.22) to meet Eq. (6.26). The range of the load R_L is 7–80 Ω, other system parameters are shown in Table 4.2. The prototype is shown in Fig. 6.30.

$$\frac{-1}{\omega C_{P1}} + \omega L_{PL} + Z_{re1\_imag} > 0 \tag{6.26}$$

Figure 6.31 shows the experimental results of θ_{re} and Z_{re} of the improved LC-filter rectifier. The θ_{re} of the improved LC filter rectifier is greater than that of the traditional rectifier. When $R_L = 7$ Ω, θ_{re} changes from –16.1° to 4.0°. The improved rectifier's 4° phase angle is designed to achieve ZVS and improve the transmission efficiency of LCT. The –16.1° phase angle of traditional rectifiers is not only detrimental to the implementation of ZVS, but also leads to a decrease in the transmission efficiency of LCT. The value of C_{P1} needs to be increased to achieve ZVS, but the low transmission efficiency of the LCT cannot be improved by changing the compensation parameters.

In addition, there is a small error between the experimental results and the simulation results. There are two main reasons. First, the parasitic inductance from the

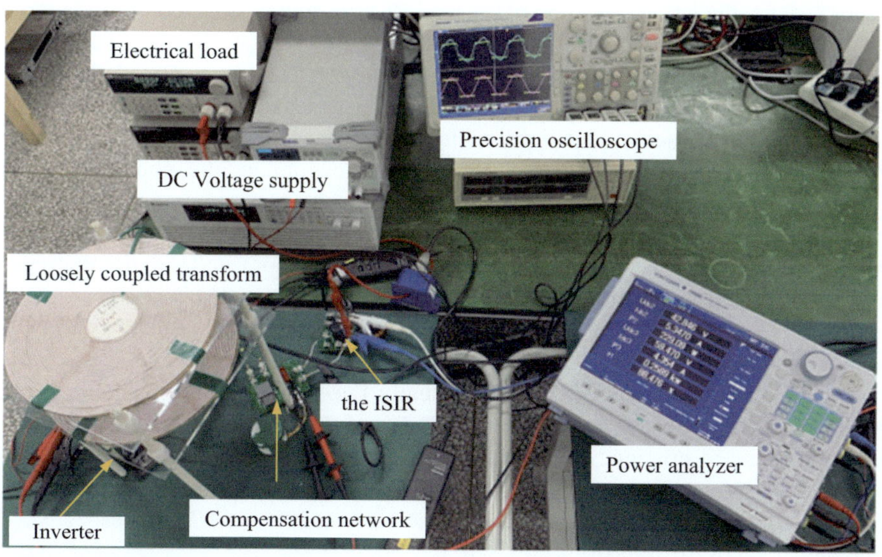

Fig. 6.30 IPT prototype with traditional S/SP compensation topology

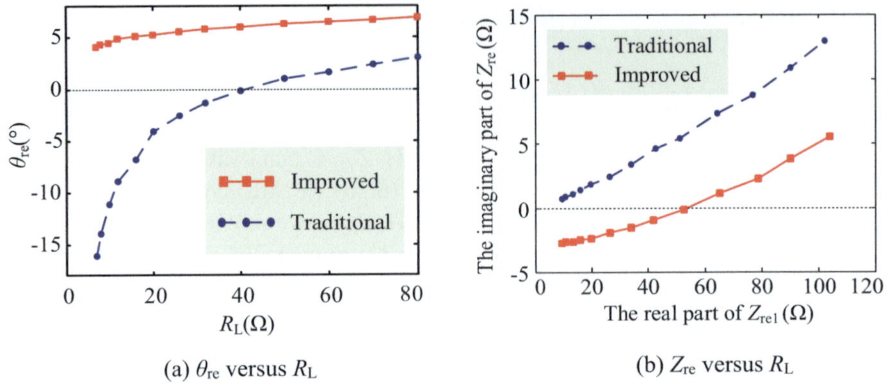

Fig. 6.31 Experimental results of θ_{re} and Z_{re} of the improved LC-filter rectifier

rectifier to the compensation capacitor C_{S2} causes L_{re} not equal to 0. From Fig. 6.26, an inductance value of 0.1 μH also has a large impact on θ_{re}. Secondly, the value of the filter inductor L_F in the experiment is smaller, which leads to fluctuations in the current of the filter inductor. The circuit analysis model is shown in Fig. 6.32a. The input voltage U_F of the filter circuit is half a wave with twice the switching frequency. The traditional S/SP compensation topology is a constant voltage output, and the input voltage U_F of the filter circuit is basically constant. The value of the filter capacitor C_F is relatively large and can be regarded as a short circuit under high frequency conditions. Therefore, the AC component of i_{LF_AC} is mainly determined by L_F, and its value is basically constant. The system has a constant voltage output characteristic, the larger the value of the load R_L, the smaller the DC current flowing through the L_F, and the greater the influence of i_{LF_AC} on the impedance angle θ_{re}. Therefore, when the load is equal to 7 Ω, the calculation results are in good agreement with the experimental results. However, as the load resistance increases, the error value of the experimental and simulation results increases. When $R_L = 80$ Ω, the waveforms of the output voltage and current of the inverter and the input voltage and current of the rectifier in the improved rectifier system are shown in Fig. 6.32b. The alternating current of the filter inductor L_F causes the phase lag of the fundamental wave of the i_{re}, and the phase difference between the input voltage and the input current of the rectifier increases, and the inductance of Z_{re} increases.

Figure 6.33 shows the curves of the system output voltage and efficiency varying with load resistance. Over the full load range, the system output voltage of the improved and traditional rectifiers is basically the same. Therefore, their output power is basically the same, but the efficiency of the improved system is higher. The difference in efficiency between the two systems is more pronounced under heavy loads. When the load resistance is equal to 7 Ω, the output power of the IPT system with the improved and traditional rectifiers is 295 and 290 W, respectively. The efficiency of the improved.

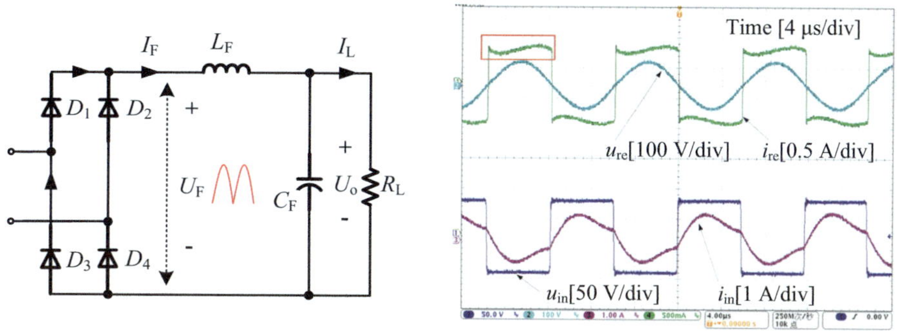

(a) Circuit analysis model (b) Experimental waveform

Fig. 6.32 Circuit analysis model and experimental waveform of the inverter and rectifier

Fig. 6.33 Efficiency and output voltage of the system versus R_L

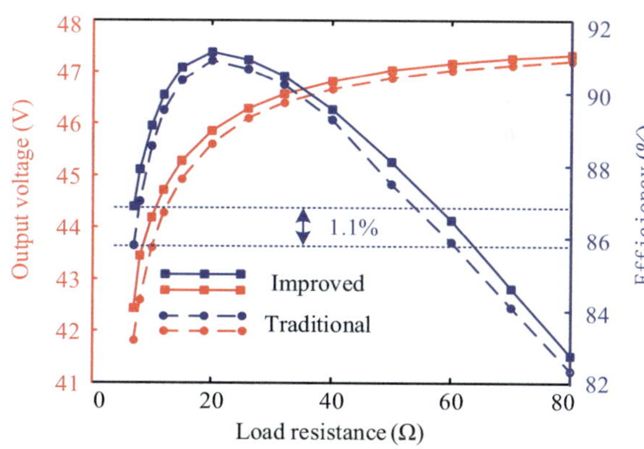

system is 1.1% higher than that of the traditional system. The comparison results prove the low loss characteristics of the improved system.

Under rated load conditions, the output voltage and current of the inverter and the input voltage and current waveforms of the rectifier are shown in Fig. 6.34. The inverter output voltage of both systems is ahead of the current, and the system input impedance is weakly inductive, which is conducive to the realization of ZVS. In the IPT system with traditional rectifier, the distortion of the rectifier input voltage causes the phase of the u_{re} to lag behind the i_{re}, and the LEACI is capacitive. In the system of improved rectifier, u_{re} is non-zero in the i region of Fig. 6.34 due to L_{re}. u_{re} is ahead of i_{re}, and the load equivalent AC impedance is weakly inductive.

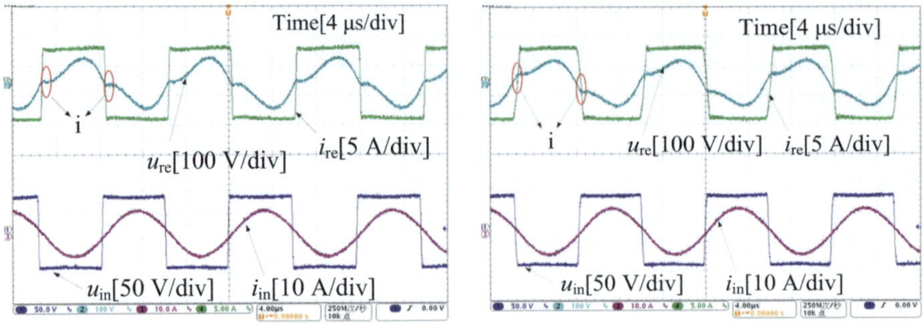

(a) Traditional IPT systems (b) Improved IPT systems

Fig. 6.34 Inverter output voltage/current and rectifier input voltage/current

6.3 Conclusion

For C-filtering and LC-filtered bridge rectifier circuits in IPT systems, the research in this chapter focuses on reactive power characteristics and improved design, including the following:

(1) The load equivalent impedance characteristics of traditional C-filter rectifiers are analyzed, and the calculation method of critical load resistance is given. The critical modal load impedance is proportional to the high-frequency output inductance.
(2) An improved C-filter rectifier structure is proposed, a small capacitor is connected in parallel in front of the traditional C-filter rectifier, and the corresponding analysis method of the LEACI is given. According to the proposed calculation method, LEACI of approximate resistance can be obtained, the reactive power is reduced, and the heavy-load efficiency of the system is increased by 0.6%.
(3) The load equivalent impedance characteristics of traditional LC-filter rectifiers are analyzed, and the calculation method of critical load resistance is given. The critical modal load impedance is inversely proportional to the high-frequency output capacitance.
(4) An improved LC-filter rectifier structure is proposed, a small inductor is connected in series in front of the traditional LC-filter rectifier, and the corresponding analysis method of the LEACI is given. According to the proposed calculation method, LEACI of approximate resistance can be obtained, the reactive power is reduced, and the heavy-load efficiency of the system is increased by 1.1%.

References

1. Wang Y, Wang H, Liang T et al (2017) Analysis and design of an LCC/S compensated resonant converter for inductively coupled power transfer. In: 2017 IEEE transportation electrification conference and E2po, Asia-Pacific (ITEC Asia-Pacific), pp 1–5
2. Wang Y, Yao Y, Liu X et al (2018) An LC/S compensation topology and coil design technique for wireless power transfer. IEEE Trans Power Electron 33(3):2007–2025
3. Li S, Li W, Deng J et al (2015) A double-sided LCC compensation network and its tuning method for wireless power transfer. IEEE Trans Veh Technol 64(6):2261–2273
4. Qu X, Han H, Wong S et al (2015) Hybrid IPT topologies with constant current or constant voltage output for battery charging applications. IEEE Trans Power Electron 30(11):6329–6337
5. Mai R, Chen Y, Zhang Y et al (2018) Optimization of the passive components for an S-LCC topology-based WPT system for charging massive electric bicycles. IEEE Trans Industr Electron 65(7):5497–5508

MIX
Papier aus verantwortungsvollen Quellen
Paper from responsible sources
FSC® C105338

If you have any concerns about our products,
you can contact us on
ProductSafety@springernature.com

In case Publisher is established outside the EU,
the EU authorized representative is:
Springer Nature Customer Service Center GmbH
Europaplatz 3, 69115 Heidelberg, Germany

Printed by Libri Plureos GmbH
in Hamburg, Germany